AF094736

Recent Advances and Future Trends in Nanophotonics

Recent Advances and Future Trends in Nanophotonics

Editors

Maria Antonietta Ferrara
Principia Dardano

MDPI • Basel • Beijing • Wuhan • Barcelona • Belgrade • Manchester • Tokyo • Cluj • Tianjin

Editors
Maria Antonietta Ferrara
Institute of Applied Sciences
and Intelligent Systems
Italy

Principia Dardano
Institute of Applied Sciences
and Intelligent Systems
Italy

Editorial Office
MDPI
St. Alban-Anlage 66
4052 Basel, Switzerland

This is a reprint of articles from the Special Issue published online in the open access journal *Applied Sciences* (ISSN 2076-3417) (available at: https://www.mdpi.com/journal/applsci/special_issues/Advances_Trends_Nanophotonics).

For citation purposes, cite each article independently as indicated on the article page online and as indicated below:

LastName, A.A.; LastName, B.B.; LastName, C.C. Article Title. *Journal Name* **Year**, *Volume Number*, Page Range.

ISBN 978-3-0365-3038-3 (Hbk)
ISBN 978-3-0365-3039-0 (PDF)

© 2021 by the authors. Articles in this book are Open Access and distributed under the Creative Commons Attribution (CC BY) license, which allows users to download, copy and build upon published articles, as long as the author and publisher are properly credited, which ensures maximum dissemination and a wider impact of our publications.

The book as a whole is distributed by MDPI under the terms and conditions of the Creative Commons license CC BY-NC-ND.

Contents

About the Editors vii

Maria Antonietta Ferrara and Principia Dardano
Special Issue on Recent Advances and Future Trends in Nanophotonics
Reprinted from: *Appl. Sci.* **2022**, *12*, 663, doi:10.3390/app12020663 1

Sanjay Keshri, Brian Rogers, Kevin Murphy, Kevin Reynolds, Izabela Naydenova and Suzanne Martin
Development and Testing of a Dual-Wavelength Sensitive Photopolymer Layer for Applications in Stacking of HOE Lenses
Reprinted from: *Appl. Sci.* **2021**, *11*, 5564, doi:10.3390/app11125564 5

Guofeng Li, Junbo Yang, Zhaojian Zhang, Kui Wen, Yuyu Tao, Yunxin Han and Zhenrong Zhang
Double Spectral Electromagnetically Induced Transparency Based on Double-Bar Dielectric Grating and Its Sensor Application
Reprinted from: *Appl. Sci.* **2020**, *10*, 3033, doi:10.3390/app10093033 23

Figen Ece Demirer, Chris van den Bomen, Reinoud Lavrijsen, Jos J. G. M. van der Tol and Bert Koopmans
Design and Modelling of a Novel Integrated Photonic Device for Nano-Scale Magnetic Memory Reading
Reprinted from: *Appl. Sci.* **2020**, *10*, 8267, doi:10.3390/app10228267 33

Teresa Crisci, Luigi Moretti and Maurizio Casalino
Theoretical Investigation of Responsivity/NEP Trade-off in NIR Graphene/Semiconductor Schottky Photodetectors Operating at Room Temperature
Reprinted from: *Appl. Sci.* **2021**, *11*, 3398, doi:10.3390/app11083398 43

Giuseppe Coppola and Maria Antonietta Ferrara
Polarization-Sensitive Digital Holographic Imaging for Characterization of Microscopic Samples: Recent Advances and Perspectives
Reprinted from: *Appl. Sci.* **2020**, *10*, 4520, doi:10.3390/app10134520 55

Luigi Sirleto, Rajeev Ranjan and Maria Antonietta Ferrara
Analysis of Pulses Bandwidth and Spectral Resolution in Femtosecond Stimulated Raman Scattering Microscopy
Reprinted from: *Appl. Sci.* **2021**, *11*, 3903, doi:10.3390/app11093903 77

Murad Althobaiti and Ibraheem Al-Naib
Recent Developments in Instrumentation of Functional Near-Infrared Spectroscopy Systems
Reprinted from: *Appl. Sci.* **2020**, *10*, 6522, doi:10.3390/app10186522 87

Zhihua Tu, Daru Chen, Hao Hu, Shiming Gao and Xiaowei Guan
Characterization and Optimal Design of Silicon-Rich Nitride Nonlinear Waveguides for 2 µm Wavelength Band
Reprinted from: *Appl. Sci.* **2020**, *10*, 8087, doi:10.3390/app10228087 113

Kaiyang Cheng, Yuancheng Fan, Weixuan Zhang, Yubin Gong, Shen Fei and Hongqiang Li
Optical Realization of Wave-Based Analog Computing with Metamaterials
Reprinted from: *Appl. Sci.* **2020**, *11*, 141, doi:10.3390/app11010141 125

About the Editors

Maria Antonietta Ferrara (Dr.) received her M.S. degree in electronic engineering from the University of Naples Federico II, Italy, in 2003 and her PhD degree in electronic engineering from the University "Mediterranea" of Reggio Calabria, Italy, in 2008. Since 2008, she has been a postdoctoral fellow at the Institute for Microelectronics and Microsystems, Naples, where she carried out research on "advanced optoelectronic devices and integrated sources based on nonlinear Raman scattering" and "noncontact optical characterization based on interferometric techniques". In 2012, she became a researcher at the Institute for Microelectronics and Microsystems, and since 2019, she has been affiliated with the Institute of Applied Sciences and Intelligent Systems of the National Research Council of Italy. She is the person in charge for the Imaging and Holography Laboratory. Her research interests are in the fields of digital holography, volume holographic optical elements, nonlinear optics, coherent Raman spectroscopy, and microscopy. She is the author of more than 90 peer-reviewed scientific papers published in indexed journals and 5 book chapters and is the holder of several patents.

Principia Dardano (Dr.) received her PhD in fundamental and applied physics in 2008. Since December 2004, she has been working at the Italian National Council of Research (CNR), first at the Institute for Microelectronics and Microsystems (IMM—CNR) and now at the Institute of Applied Sciences and Intelligent Systems (ISASI), in Naples, where she is the head of the photolithographic laboratory. She is responsible for the design, fabrication, and characterization of theranostic and photonic devices and for the imaging of micro/nano biosystems with AFM, EFM, and SEM.

Editorial
Special Issue on Recent Advances and Future Trends in Nanophotonics

Maria Antonietta Ferrara * and Principia Dardano *

National Research Council (CNR), Institute of Applied Sciences and Intelligent Systems, Via Pietro Castellino 111, 80131 Naples, Italy
* Correspondence: antonella.ferrara@na.isasi.cnr.it (M.A.F.); principia.dardano@na.isasi.cnr.it (P.D.)

Citation: Ferrara, M.A.; Dardano, P. Special Issue on Recent Advances and Future Trends in Nanophotonics. *Appl. Sci.* **2022**, *12*, 663. https://doi.org/10.3390/app12020663

Received: 5 January 2022
Accepted: 6 January 2022
Published: 11 January 2022

Publisher's Note: MDPI stays neutral with regard to jurisdictional claims in published maps and institutional affiliations.

Copyright: © 2022 by the authors. Licensee MDPI, Basel, Switzerland. This article is an open access article distributed under the terms and conditions of the Creative Commons Attribution (CC BY) license (https://creativecommons.org/licenses/by/4.0/).

1. Introduction

Nanophotonics is an emerging multidisciplinary frontier of science and engineering. Its high potential in contributing to the development of many areas of technology makes nanophotonics a focus of interest for many researchers from different fields.

In light of the above, the present Special Issue of *Applied Sciences* on "Recent Advances and Future Trends in Nanophotonics" gives an overview of the latest developments in nanophotonics and its roles in different application domains. Eleven papers were submitted to this Special Issue, and nine papers were accepted (i.e., an 82% acceptance rate). The presented papers explore innovative trends of nanophotonic science and technology that enable technological breakthroughs in high-impact areas and cover several topics, mainly regarding diffraction elements, detection, imaging, spectroscopy, optical communications and computing.

2. Diffraction Elements

Diffractive Optical Elements (DOEs) have been expanding for many years and are a stimulant technology that allow for the redirection of redirect light using diffraction rather than refraction. When a photonic structure is realized via holography, we refer to this as Holographic Optical Elements (HOEs), i.e., elements recorded by a specific interference pattern in a photosensitive optical material. Volume HOEs show very high diffraction efficiency in single-order diffraction; however, angular wavelength selectivity and high dispersion still are considerable challenges. In order to achieve a highly efficient HOE with low dispersion, Keshri and co-workers [1] explored a fabrication method regarding a DOE with two separate target wavelengths. Two independently sensitized layers of photopolymer were stacked and used at two different wavelengths in sequential holographic recording. The diffraction peaks of the recorded photonic structures were investigated and compared with those of the designed structures, and then, they were theoretically predicted. The device, illuminated with an expanded divergent beam at both target wavelengths and with white light, showed a strong diffracted beam. This approach could be optimized and useful for potential applications in the fields of illumination, solar collection and displays.

A diffraction grating structure was numerically simulated in the paper authored by Li and coworkers [2], where a guided-mode resonance system with a double-bar dielectric grating was exploited to achieve a multispectral electromagnetically induced transparency (EIT) effect. In this case, the resonance wavelengths of the two EIT peaks could be modified by changing the corresponding structural parameters and could be exploited for dual mode sensor, dual channel slow light, etc. When optimized, two EIT peaks showed ultra-high Q factors of 35,104 and 24,423. Moreover, the dual-mode refractive index sensor based on this system reached figures of merit (FOMs) of 571.88 and 587.42.

3. Detection, Imaging and Spectroscopy

The manipulation of light and light-matter interactions have been largely investigated for applications in the fields of detection, imaging and spectroscopy.

In the paper published by Demirer and co-workers [3], the Magneto-Optic Kerr Effect in the guided mode of an unbalanced Mach Zehnder Interferometer (MZI) was designed and simulated to optically detect the magnetization direction of ultra-thin (~12 nm) metal cladding. In fact, the device was an unbalanced MZI based on InP membrane on silicon. The MZI arms were made up of a polarization converter from one side and ferromagnetic thin-film cladding and a delay line from the other side. The device read a nanoscale memory bit (400 nm × 50 nm × 12 nm) with a signal-to-noise ratio ~10 dB and tolerated performance reductions that arose during the fabrication. While this hybrid device based on ultra-thin metal membrane on silicon was demonstrated to be an all-optical magnetic memory reading tool, hybrid devices with an ultra-thin conductive layer jointed with semiconductors were presented by Crisci and co-authors [4], such as all-optical Schottky photodetectors that operated at room temperature. In particular, in [4], two Schottky photodetectors based on graphene/n-silicon (Si) and graphene/n-germanium (Ge) Schottky barriers were theoretically investigated and operated at 1550 nm and at 2000 nm, respectively. The responsivity/noise equivalent power (NEP) ratio was analysed, and a strong addiction on the Schottky barrier height of the junction was demonstrated. The authors derived a closed analytical formula for use in maximizing the responsivity/NEP ratio and theoretically discussed how the Schottky barrier height is related to the reverse bias applied to the junction. Moreover, they found that at 1550 nm, the optimized graphene/n-silicon (Si) Schottky PDs with a reverse bias of 0.66 V showed a responsivity and NEP of 133 mA/W and 500 fW/$\sqrt{Hz}$, respectively. Finally, at 2000 nm, the optimized graphene/n-germanium (Ge) Schottky PDs showed a responsivity and NEP of 233 mA/W and 31 pW/$\sqrt{Hz}$, respectively.

The detection and imaging of optical signals were investigated and reviewed in this Special Issue. In particular, two innovative imaging techniques were addressed: Polarization-sensitive Digital Holographic Imaging (PDHI) [5] and Stimulated Raman Scattering (SRS) microscopy [6].

Coppola and Ferrara [5] reported a review on the state of the art of PDHI techniques, focusing on the theoretical principles and important applications. The paper not only provided an exhaustive review of applications in several fields, from biology to microelectronics and micro-photonics, but also emphasized the merits of this new technique based on the interference between different polarized optical beams. PSDHI, in fact, simultaneously allows the three-dimensional reconstruction and the quantitative evaluation of the polarization properties of a sample with a resolution on the micrometric scale, a good acquisition speed and the absence of labels/markers.

Sirleto and co-workers [6] measured the spectral resolution of SRS, i.e., the ability to distinguish closely lying resonances, and their paper focused on the spectral splitting of protein and lipid bands in the C-H region, which is of great interest in the field of biochemistry. In particular, the paper addressed the interplay among pump and Stokes bandwidth and the degree of chirp-matching. Moreover, the spectral resolution of femtosecond SRS microscopy was experimentally investigated.

Spectroscopy systems were further reviewed by Althobaiti and Al-Naib [7] in the field of instrumentation working at near-infrared NIR frequencies. The authors discussed NIR spectroscopy systems from the instrumentation point of view with regard to state-of-the-art approaches and the associated challenges. In particular, the authors provided a summary of the recent development of continuous-wave, time-domain and frequency-domain NIR systems and presented an outlook into the future of the design and development of functional near-infrared spectroscopy systems for various medical applications.

4. Optical Communications and Computing

Nonlinear waveguides can play a key role in optical communication applications, in particular in the 2 μm wavelength band, where interest in mitigating the 'capacity crunch' is growing. In this context, Tu and co-authors [8] fabricated and characterized silicon-rich nitride (SRN) ridge waveguides with different widths and rib heights. The structures showed a loss of ~2 dB/cm, and the SRN nonlinear refractive index was shown to be ~1.13×10^{-18} m^2/W around the wavelength 1950 nm. By optimizing parameters to improve nonlinear performances for the 2 µm band, a maximal nonlinear figure of merit (i.e., the ratio of nonlinearity to loss) of 0.0804 W^{-1} or a super-broad FWM bandwidth of 518 nm were found. These results pave the way for high-performance on-chip nonlinear waveguides for use in optical communications in the 2 µm wavelength band.

Finally, the review proposed by Cheng and co-authors [9] underlined how the recent development of nanofabrication technologies for the generation, processing and detection of optical signals have paved the way in the field of new analog optical computing. Particular attention was given to metamaterials or metasurfaces, which offer unprecedented opportunities to arbitrarily manipulate light waves at the subwavelength scale, leading to an acceleration of the progress of wave-based spatial analog computing. Furthermore, the authors discussed challenges and future opportunities in high-efficiency signal processing by exploiting quantum behaviors.

Funding: This research received no external funding.

Acknowledgments: This issue would not be possible without the valuable contributions of all the authors and peer reviewers. We would like to take this opportunity to record our sincere gratefulness to the editorial team of *Applied Sciences*.

Conflicts of Interest: The authors declare no conflict of interest.

References

1. Keshri, S.; Rogers, B.; Murphy, K.; Reynolds, K.; Naydenova, I.; Martin, S. Development and Testing of a Dual-Wavelength Sensitive Photopolymer Layer for Applications in Stacking of HOE Lenses. *Appl. Sci.* **2021**, *11*, 5564. [CrossRef]
2. Li, G.; Yang, J.; Zhang, Z.; Wen, K.; Tao, Y.; Han, Y.; Zhang, Z. Double Spectral Electromagnetically Induced Transparency Based on Double-Bar Dielectric Grating and Its Sensor Application. *Appl. Sci.* **2020**, *10*, 3033. [CrossRef]
3. Demirer, F.E.; van den Bomen, C.; Lavrijsen, R.; van der Tol, J.J.G.M.; Koopmans, B. Design and Modelling of a Novel Integrated Photonic Device for Nano-Scale Magnetic Memory Reading. *Appl. Sci.* **2020**, *10*, 8267. [CrossRef]
4. Crisci, T.; Moretti, L.; Casalino, M. Theoretical Investigation of Responsivity/NEP Trade-off in NIR Graphene/Semiconductor Schottky Photodetectors Operating at Room Temperature. *Appl. Sci.* **2021**, *11*, 3398. [CrossRef]
5. Coppola, G.; Ferrara, M.A. Polarization-Sensitive Digital Holographic Imaging for Characterization of Microscopic Samples: Recent Advances and Perspectives. *Appl. Sci.* **2020**, *10*, 4520. [CrossRef]
6. Sirleto, L.; Ranjan, R.; Ferrara, M.A. Analysis of Pulses Bandwidth and Spectral Resolution in Femtosecond Stimulated Raman Scattering Microscopy. *Appl. Sci.* **2021**, *11*, 3903. [CrossRef]
7. Althobaiti, M.; Al-Naib, I. Recent Developments in Instrumentation of Functional Near-Infrared Spectroscopy Systems. *Appl. Sci.* **2020**, *10*, 6522. [CrossRef]
8. Tu, Z.; Chen, D.; Hu, H.; Gao, S.; Guan, X. Characterization and Optimal Design of Silicon-Rich Nitride Nonlinear Waveguides for 2 µm Wavelength Band. *Appl. Sci.* **2020**, *10*, 8087. [CrossRef]
9. Cheng, K.; Fan, Y.; Zhang, W.; Gong, Y.; Fei, S.; Li, H. Optical Realization of Wave-Based Analog Computing with Metamaterials. *Appl. Sci.* **2021**, *11*, 141. [CrossRef]

Article

Development and Testing of a Dual-Wavelength Sensitive Photopolymer Layer for Applications in Stacking of HOE Lenses

Sanjay Keshri [1], Brian Rogers [1,2], Kevin Murphy [1], Kevin Reynolds [1,2], Izabela Naydenova [1,2] and Suzanne Martin [1,*]

[1] Centre for Industrial and Engineering Optics, Dublin City Campus, FOCAS Research Institute, School of Physics and Clinical & Optometric Sciences, College of Science and Health, Technological University Dublin, 13 Camden Row, D08 CKP1 Dublin 8, Ireland; sanjayceloscusat@gmail.com (S.K.); brian.rogers@tudublin.ie (B.R.); kevin.p.murphy@tudublin.ie (K.M.); C17510763@mytudublin.ie (K.R.); izabela.naydenova@tudublin.ie (I.N.)
[2] School of Physics and Clinical & Optometric Sciences, College of Science and Health, Dublin City Campus, Technological University Dublin, Central Quad, Grangegorman Lower, D07 ADY7 Dublin, Ireland
* Correspondence: suzanne.martin@tudublin.ie

Citation: Keshri, S.; Rogers, B.; Murphy, K.; Reynolds, K.; Naydenova, I.; Martin, S. Development and Testing of a Dual-Wavelength Sensitive Photopolymer Layer for Applications in Stacking of HOE Lenses. *Appl. Sci.* **2021**, *11*, 5564. https://doi.org/10.3390/app11125564

Academic Editors: Maria Antonietta Ferrara and Principia Dardano

Received: 7 May 2021
Accepted: 10 June 2021
Published: 16 June 2021

Publisher's Note: MDPI stays neutral with regard to jurisdictional claims in published maps and institutional affiliations.

Copyright: © 2021 by the authors. Licensee MDPI, Basel, Switzerland. This article is an open access article distributed under the terms and conditions of the Creative Commons Attribution (CC BY) license (https://creativecommons.org/licenses/by/4.0/).

Abstract: Diffractive optical elements (DOEs) have been in development for many years and are an exciting technology with the capability to re-direct light, using diffraction rather than refraction. Holographic Optical Elements (HOEs) are a subset of diffractive optical elements for which the photonic structure is created holographically, i.e., by recording a specific interference pattern in a suitable, photosensitive optical material. Volume HOEs are of particular interest for some applications because of their very high diffraction efficiency and single diffracted order; however, high dispersion and angular wavelength selectivity still present significant challenges. This paper explores a method for producing a compound DOE useful for situations where elements designed for two separate target wavelengths can be advantageously combined to achieve a highly efficient HOE with reduced dispersion. A photopolymer material consisting of two independently sensitized laminated layers is prepared and used in sequential holographic recording at two different wavelengths. The photonic structures recorded are investigated through examination of their diffraction peaks and comparison with the structure predicted by modeling. Finally, the device is illuminated with an expanded diverging beam at both target wavelengths and with white light, and a strong diffracted beam is observed.

Keywords: diffractive elements; holography; photopolymer; holographic optical elements; volume gratings; compound optical elements; stacked gratings

1. Introduction

Diffractive Optical Elements (DOEs) are photonics structures that focus and redirect light, using diffraction rather than refraction. 'Volume' or 'thick' Diffractive Optical Elements utilize Bragg diffraction and can very efficiently transfer energy from the incident beam into a diffracted beam [1]. DOEs have the potential to replace optical components in a range of applications and have the advantages of being low cost, lightweight and in most cases, much thinner than their refractive counterparts.

The DOE element can be thought of as a set of localized gratings, each potentially having a different slant, orientation and spatial frequency (set by the patterning process) so that the direction and divergence of the diffracted output beam can be chosen as desired [2]. As a result, DOEs can replace standard optical components (e.g., lenses and mirrors) in some applications. Bragg diffraction, however, introduces wavelength and angular selectivity, which can be particularly restrictive if the element/grating is thick [1].

Photopolymer materials are typically prepared as thin, photosensitive films that polymerize when they absorb light, and can record patterns with very high spatial resolution. They can, therefore, be used to form volume DOEs in a low-intensity, single-exposure step, which is readily customized. Since the patterning step involves holographic recording, they are also often referred to as Holographic Optical Elements or HOEs. Currently, Diffractive Optical Elements are not routinely used with broadband sources. Some practical applications utilizing surface gratings in the management of light are reported in the literature; for example, LED light is re-directed through a 30° angle by a binary diffractive lens [3] and broadband multicolor imaging is demonstrated by using multi-level surface holograms [4].

Fabrication of surface gratings is challenging, however, since it requires multiple manufacturing steps—lithography and etching—and the surface of the photonic structures is difficult to protect. Volume photopolymer HOEs [5] offer a significant improvement since they can be patterned (with low intensity exposure) [6] through the volume of the materials, can be laminated [7,8] into a stack, and importantly, they can achieve near 100% re-direction of the incident light into a single output beam [9,10]. This optical patterning can potentially replace the etching and stamping process and facilitate customization and small batches. The main features of surface and volume HOEs are compared, at a glance, in Table 1.

Table 1. Comparing surface and volume diffractive optics elements.

	Surface DOEs	Volume DOEs
Fabrication	Multiple steps, including wet processing	Single step (self-processing photopolymers)
Protection and Stacking	More difficult since the pattern is on the surface	Easier, the pattern is in the volume
Diffraction Efficiency	For sinusoidal profiles theoretical limit 33%	Up to 100%

Research into beam patterning and shaping with HOEs has continued to advance [11–13], but solutions for broad wavelength sources are needed to increase the application range. Many sources, such as white light LEDs, output a set of wavelength ranges rather than a continuous broad band of wavelengths (as for solar radiation, for example). This paper explores a method for producing a compound element useful for situations where elements that are designed for two separate target wavelengths can be advantageously combined to achieve a highly efficient HOE. The authors previously reported stacked gratings for operation in broader wavelength range by calculating the correct angle of diffraction for each target wavelength and recording at a single wavelength in green in two separate layers that are then stacked after recording [14]. This approach is somewhat limited by the complex modeling and recording arrangements needed for elements such as lenses, as they contain a wide range of spatial frequencies and slants that need to be matched to the different wavelength, as well as the difficulty of assembling the stack after recording without introducing damage and/or shifts in the slant angle.

This paper presents a method of stacking that utilizes selectively pre-sensitized layers stacked together so that dual wavelength recording can take place independently in the two stacked layers and explores the process as a means to combine lens elements. The recordings are independent of one another because the dye sensitizer in each layer is matched to one of the recording laser wavelengths. It is demonstrated here for two layers but has the potential to be developed for additional layers (e.g., red, green, blue).

The stacking/multiplexing approach to correcting for chromatic dispersion is conceptually simple and relies on fabrication and combination of devices tailored to specific wavelength ranges. It has been exploited to multiplex or de-multiplex in diffusers [15] and lenses [16], in stacking meta-surfaces [17], and more recently, to wavelength multiplexing in lens arrays for full color imaging [18] with each layer fabricated of different materials and with different design parameters to optimize it for a specific frequency band. However, these devices are fabricated through lithographic processes or holographically recorded into a single layer of a pan-chromatic material through multiplexing. Here, we utilize the limited wavelength sensitivity range in the pre-stacked photopolymer layers to exploit the simple full field exposure step to holographically record a compound element at two separate wavelengths, without the need for complex modeling and fabrication techniques.

In the literature, a variation of dye concentration in multilayer photopolymer was used to improve the uniformity in volume gratings [19]. Three dyes were combined in a single layer of photopolymer material for full color display holography [20]. In this work, two photopolymer layers sensitive to two different wavelengths were laminated together and a lens was recorded in each layer by their designed wavelength in sequential manner.

Firstly, a blue-sensitive photopolymer formulation, previously developed for the recording of reflection holograms (at high spatial frequency), was optimized for the low spatial frequency and high layer thickness needed in the proposed off-axis collimating lens transmission-format element. Next, the design considerations for stacked lenses were discussed for the application of off-axis, high-diffraction efficiency collimating HOE elements. Finally, the dual-wavelength sensitive multi-layer structure was prepared by stacking two photopolymer layers and used to successfully record collimating elements for blue and green wavelengths. The local diffraction characteristics of the recorded microstructure were tested and limitations discussed.

1.1. Concept

Figure 1 shows the basic concept for stacking by recording in a dual-wavelength sensitive multi-layer structure. Holographic recording of lenses is achieved through the interference of two beams that mimic the input and output beams of the lens device. The interference pattern created is recorded in the volume of the photosensitive material as a periodic structure in which the spatial frequency and slant at a particular point are defined by the two wave-vectors normal to the interfering wave fronts at that point. This structure diffracts incoming light of the same wavelength and incident angle to re-create the original beams. In this case, a dual layer is prepared in which each layer responds optimally to a limited range of wavelengths in blue or green. This means that two sequential recordings in blue and green create a stack of two lenses, without the need for repositioning or adjustment of the layer. Due to the high selectivity of volume diffractive lenses, each lens reconstructs only the appropriate wavelength range and the output beams are identical for both wavelengths. This approach has potential in reducing chromatic dispersion in HOE lenses.

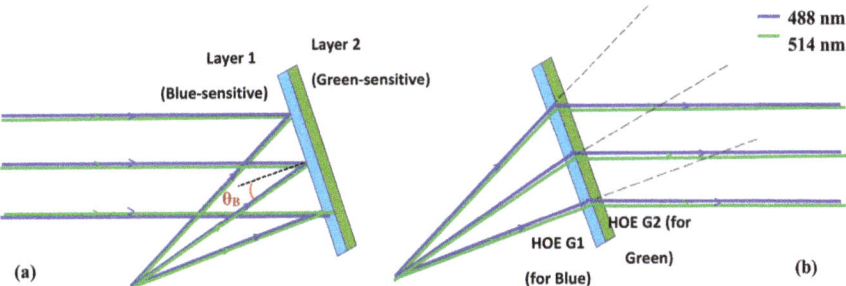

Figure 1. Schematic diagram for (**a**) recording two lens patterns, using two different wavelengths in a dual-sensitive layer to create stacked lenses. (**b**) Replay process showing that the stacked lenses reproduce the same collimated beam for both input wavelengths.

1.2. Theory

The characteristics of a beam diffracted from a grating depend not only on its orientation and spatial frequency (photonic structure), but also on the angles and wavelength of the incident beam.

The classical grating equation for transmission [21] can be expressed by Equation (1):

$$k_i \sin\theta_0 + k_d \sin\theta_{+1} = \frac{2\pi m}{n\Lambda} \qquad (1)$$

where, $k_i = \frac{2\pi}{\lambda_0}$ and $k_d = \frac{2\pi}{\lambda_1}$ are the propagation vector's modulus for the incident and diffracted beams, m = order of diffraction (m = 1 for volume HOE), θ_0 = the angle of incidence in air, θ_{+1} = angle of diffraction in air, n = refractive index of the medium, and Λ = period of the refractive index modulation (spatial period).

Volume HOEs work on the principle of Bragg's law [22], and at the Bragg condition, the angle of incidence equals the angle of diffraction ($\theta_{+1} = -\theta_0$ and $\lambda_0 = \lambda_1$). Equation (2) defines the wavelength at the Bragg condition.

$$\lambda = 2n\Lambda \sin\left(\frac{\theta_{+1} - \theta_0}{2}\right) \qquad (2)$$

In applications where the re-direction of illumination is important, volume diffractive optics are preferred because of their potential for high diffraction efficiency, meaning that a large proportion of the incident light is diffracted into a single diffraction order in a direction defined by 2. The diffraction efficiency achieved also depends on the material characteristics of the grating, such as refractive index modulation and thickness. The Kogelnik Coupled Wave Theory (KCWT) [1] is a well-established tool for modeling the efficiency of such elements and can predict how the efficiency of Volume HOEs depends on many parameters, such as the illuminating beam incidence angle, replay wavelength, index modulation and grating thickness.

KCWT shows that at Bragg incidence, the diffraction efficiency for the volume grating can be modeled using the following:

$$\eta = \mathrm{Sin}^2\left(\frac{\pi n_1 d}{\lambda \mathrm{Cos}\theta_B}\right) \qquad (3)$$

where θ_B = Bragg angle inside the layer, d is the thickness of the grating and n_1 the refractive index modulation.

The Q factor is used to check whether the grating will function as a volume Bragg grating [23].

$$Q = \frac{2\pi \lambda d}{n\Lambda^2} \qquad (4)$$

With the parameters defined as above, a Q factor of > 10 is considered to be in the 'volume' regime, and therefore is expected to have a single diffraction order and behavior that can be predicted using KCWT.

2. Materials and Methods

The dual-sensitive recording approach was explored here through a combination of modeling, experimental work, materials optimization, fabrication and testing. Firstly, the approximate diffractive lens device requirements (spectral range, thickness, spatial frequency range, refractive index modulation) for each layer were determined though modeling. Then, a previously developed blue-sensitive photopolymer formulation was optimized for the spatial frequency range and thickness needed here. This was achieved through a study of gratings holographically recorded in dry layers coated on glass slides and optically patterned using different exposure energies and inter-beam angles in the overlapping beams. Finally, holographic lens elements were recorded, using both blue (488 nm) and green (514 nm) wavelengths separately and together (in the dual-sensitive layer stack), and the resulting HOEs were studied through the analysis of theory diffraction behavior, allowing the recorded microstructure to be studied.

2.1. Materials Preparation

Two separate photopolymer formulations were prepared. First, a polyvinyl alcohol stock solution was prepared as described elsewhere [14] (10 g dissolved into 100 mL). Two separate dye solutions were then prepared, both with 0.11 g of powdered dye (erythrosine B or acryflavin) in 100 mL of deionized water. Next, to make the green-sensitive formulation, the electron donor (triethanolamine (2 mL)) and the monomers (0.6 g acrylamide and 0.2 g methylene bisacrylamide) were added to 17.5 mL of the PVA solution and stirred for one hour. Finally, 4 mL of the erythrosine B solution was added. For the blue-sensitive formulation, triethanolamine (2 mL) and the monomers (0.8 g acrylamide and 0.2 g methylene bisacrylamide) were added to 17.5 mL of the PVA solution and stirred for one hour. Finally, 3 mL of the acryflavin solution was added.

Layers were prepared by using a micropipette to deposit a controlled amount of solution onto either glass microscope slides or high optical quality plastic of the same dimensions on a level surface. After allowing them to dry in a darkroom for 1–2 days, the dual-sensitive layer photopolymer was prepared by laminating the two layers together. For this step, it was important that one of the layers was prepared on a flexible substrate, in this case, Makrolfol of 0.375 mm thickness (flexible glass is also very suitable [14]). This allows for lamination of the two photopolymer surfaces together in a stable layer with no air gaps or bubbles.

2.2. Optical Patterning: Holographic Recording Setup

An off-axis holographic lens can be recorded by using the setup shown in Figure 2. Two collimated beams are arranged such that they overlap on the photopolymer layer, then a suitable lens is inserted into one of the beams such that it focuses the beam before it reaches the photopolymer as shown. This produces a diverging beam to interfere with the collimated reference beam and a HOE lens is recorded. The numerical aperture of the lens determines the divergence of the beam and the range of the angles of incidence at the photopolymer. This kind of recording was demonstrated previously with both cylindrical and spherical lenses [24]. However, in this instance, a cylindrical lens was used so that the beam only diverged in one axis (horizontal). This simplified the modeling and experimental testing since the grating characteristics varied in only one axis. In the blue sensitive photopolymer formulation optimization work, simple gratings were recorded, and the set-up was identical except for the fact that both beams were collimated, as the cylindrical lens was not present. For some characterization work, a 532 nm and a 473 nm laser were used but for recording in the dual-sensitized layer, an argon ion was used so that both the 514 nm and 488 nm laser lines could be used in the same optical setup.

When recording lenses, the range of angles of incidence which the diverging beam produces determines the inter-beam angles for the interference pattern, and consequently, the spatial frequency and slant at each location of the photopolymer. The cone of rays will be the same (assuming negligible aberrations) for recordings at both green and blue recording wavelengths. However, since the spatial frequency also depends on the wavelength of the incident light, this will lead to a different interference pattern for each recording.

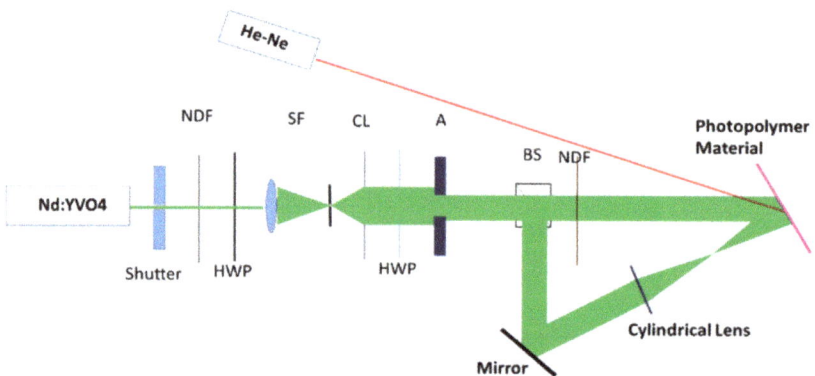

Figure 2. Experimental set-up for both the recording of a cylindrical holographic lens in green and probing using a He–Ne laser. NDF—neutral density filter; HWP—half-wave plate; SF–spatial filter; CL—collimator lens; A—variable aperture; BS—beam splitter.

2.3. Testing the Recorded Lenses

The recorded lenses were tested at three locations on the lens: at the center of the lens, 3 mm to the left of the center and 3 mm to the right of the center. Figure 3 illustrates how the Bragg positions for the two gratings vary because of the different wavelengths, and how they differ at the three probing positions. Testing involved illumination with an unexpanded He–Ne beam of about 1 mm in diameter and rotation of the sample through a wide angular range while monitoring the intensity changes in the transmitted or first-order beam. Since in all cases studied here, higher order beams are negligible, it was often more useful to monitor the zero order or transmitted beam so that the angular position, diffraction efficiency and FWHM information was available in a single scan. Except where otherwise stated, all tests were carried out in a computer-controlled system with a Newport rotation stage and photodetector and a λ = 633 nm He–Ne probe laser. Figure 3 shows an illustration of the different conditions for Bragg matching at three positions on the lens. The wave vectors for the input and output beams at both wavelengths are shown along with the grating vector for each position on the lens for each of the two wavelengths.

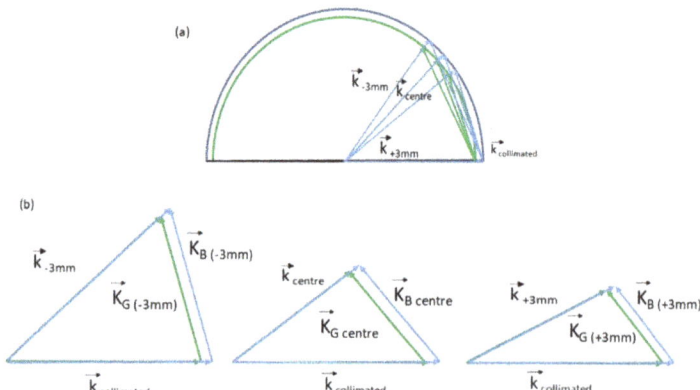

Figure 3. (a) Ewald Sphere illustration of the different conditions for Bragg matching at three positions on the lens and (b) Bragg matching condition for each position, −3 mm, center and +3 mm shown separately for clarity.

2.4. Modeling the Recorded Microstructure—Calculation of Slant and Spatial Frequency

In order to fabricate a HOE lens which collimates an off-axis diverging beam, a photonic structure is needed that will be on-Bragg everywhere for rays diverging from the off axis focal point and will re-direct them (through diffraction) into a collimated beam with a defined angle with respect to the HOE plane. At each location, the desired input beam angle can be defined, using the position of the focal point relative to that location. The desired output beam angle is known and is the same everywhere on the element (for the collimated output). Once the angles of the two beams, θ_1 and θ_2, are known, the slant angle ($\theta_S = (\theta_2 + \theta_1)/2$) and Bragg angle ($\theta_B = (\theta_2 - \theta_1)/2$) at that location are determined from geometry, and the spatial frequency needed to obtain such a Bragg angle at the design wavelength is calculated from Equation (2) above. Figure 4 depicts this procedure for a single point on the DOE lens. Once the slant angle and spatial frequency are known, one can calculate, for every point on the lens, the interference pattern required to fabricate the desired structure at a given wavelength. If the recording wavelength is close to the design wavelength, however, this process becomes extremely convenient, as θ_1 and θ_2 become the recording angles. Then the process involves using the desired input beam as one of the recording beams, made to interfere in the plane of the polymer material with the other recording beam, which has the characteristics of the desired output beam. The dual-sensitive polymer layer enables this convenient process to be exploited at two separate wavelengths in one device.

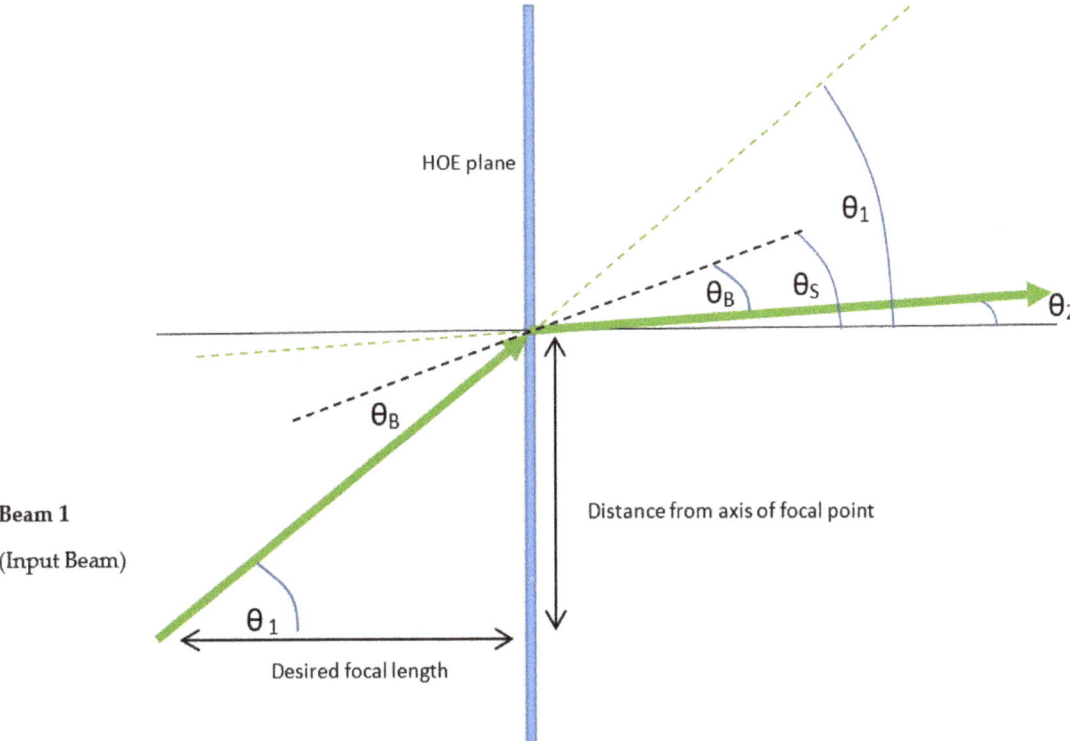

Figure 4. Simple geometrical model used to calculate the required grating slant and spatial frequency for the grating needed to re-direct Beam 1 into Beam 2 (at each specific location on the DOE lens). For simplicity, refraction at the medium boundaries are not shown here, but are taken into account in the calculations. θ_B is the Bragg angle; θ_1 is the incident angle of the input Beam 1 and θ_2 is the angle of the output Beam 2. θ_S denotes the slant angle of the fringes.

3. Results

3.1. Modeling and Design of the Diffractive Optical Elements

The stack envisaged consists of two off-axis lens elements, one designed for a blue target wavelength, and one designed for a green, both having a focal point of 5 cm from the plane of the photopolymer element.

The design considerations are as follows:

- Photonic structure: Each grating needs to have a tailored range of spatial frequency and slant to produce correct output angle (at that wavelength) for every point on the diffractive lens (Equations (1) and (2))
- Single diffracted beam: The design needs to ensure that each grating is functioning as a volume grating at all points (to avoid multiple diffracted beams) (Equation (4))
- Grating strength: Each grating must have a specific Refractive Index Modulation to produce high DE for its target wavelength (Equation (3))
- Avoiding crosstalk: Each grating must have a sufficiently narrow angular and spectral selectivity curve so that each grating will only re-direct light of the wavelength that it is designed for.

Target wavelengths [25] 514 nm and 488 nm were chosen to mimic typical green and blue peaks in a white LED spectrum for which close laser wavelengths were available for holographic recording. The design is based on high efficiency with the assumptions that higher diffracted orders are negligible, the grating is uniform though the depth of the polymer layer, and the polymer faithfully records the interference pattern (zero dimensional changes during recording).

As discussed above, each element consists of a range of grating spatial frequencies, which will change with the design wavelength, the off-axis angle and the aperture and focal length of the lens. In turn, the set of spatial frequencies recorded in the lens influences all of the above design considerations. On-axis elements are not suitable here because of well-understood issues around the limits of volume diffraction [2] that arise where the diffraction angle is very small (near zero spatial frequency). Off-axis elements are, therefore, used to focus light for these applications, so the first design step was to choose an offset angle, or central spatial frequency, that would avoid this. For transmission elements, high spatial frequencies (large offset angles) are less challenging to record so the limits of the low spatial frequency range are explored here.

3.1.1. Green Lens Design

First, the range of spatial frequencies expected to be recorded across the lens for a recording wavelength of 514 nm was calculated for a number of different off-axis angles (different central spatial frequencies). The calculated range of spatial frequencies for lenses having central spatial frequencies of 300, 400, 455 and 500 L/mm are shown in Figure 5a. Figure 5b shows the calculated Q values for the lowest spatial frequency occurring in each lens from Figure 5a for different thicknesses. It was decided to choose the lens with the lowest off-axis angle (and spatial frequency range) that would ensure volume Bragg grating behavior ($Q > 10$) across the whole lens aperture for 150 microns in thickness. This ensures that higher orders are negligible. A minimum of 140–150 micron-thick gratings were shown to be needed to avoid crosstalk in previous work [14]; thickness above 150 microns can be more challenging to prepare without compromising layer optical quality. The lens chosen was, therefore, Lens 3, with 150 μm layer thickness and 455 L/mm central spatial frequency.

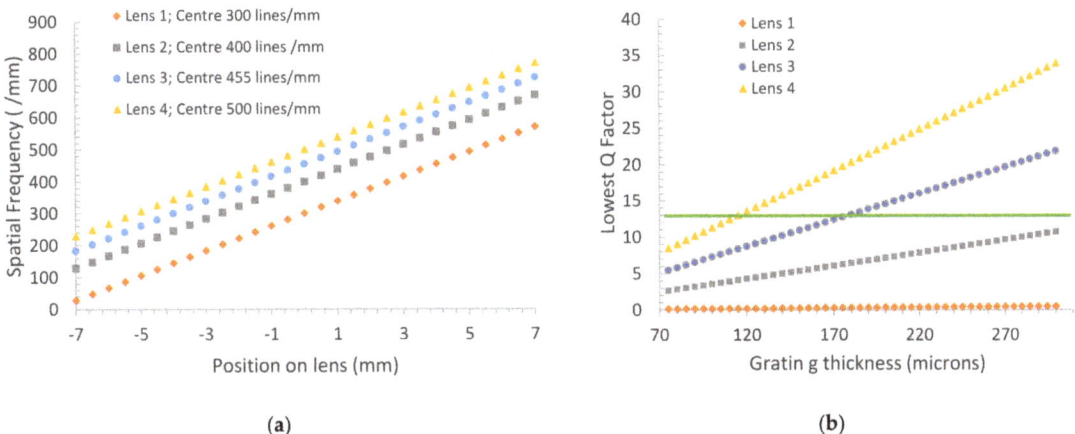

Figure 5. (a) Local grating spatial frequency across the diameter of four lenses with different central spatial frequencies, design wavelength 514 nm, focal length 5 cm, lens diameter 14 mm and (b) the Q value for the lowest spatial frequency in each lens for a range of grating thicknesses. The green horizontal line indicates where Q = 10, i.e., the minimum Q value to ensure a single diffracted order.

3.1.2. Blue Lens Design

The next step was to design a second lens, which would produce the same output angle across the whole lens aperture for a 488 nm design wavelength with the same range of input and output angles. The Q value for the lowest spatial frequency was 11.4, for a 150 micron layer, ensuring volume diffraction here, too. Since 150 micron layers were used, the capability of achieving sufficient grating strength was assured for the green sensitive formulation (100% can, in fact, be achieved in much thinner layers [26]). Figure 6 shows the range of spatial frequencies across this lens, which has the same angular offset and focal length as Lens 3 but is designed for operation at λ = 488 nm.

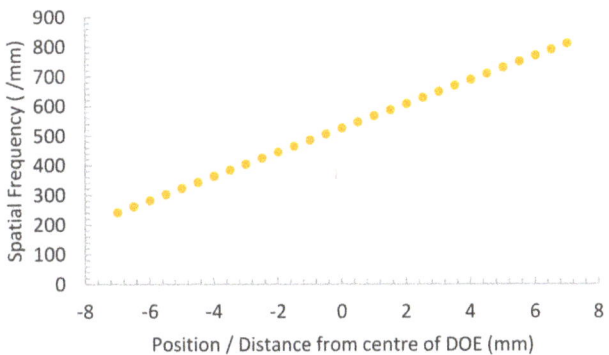

Figure 6. The range of spatial frequencies across the lens designed for 488 nm, diameter = 14 mm with the same angular offset and focal length as Lens 3 but for operation at λ = 488 nm.

The refractive index modulation needed to maximize the diffraction efficiency was controlled through exposure during recording, as detailed in the experimental section. Some characterization was needed in order to achieve this for the blue-sensitive formulation, which had not been as widely studied as the standard green (eythrosine B) formulation.

3.2. Optimization of the Acriflavine Based Photopolymer Layer

In the literature, acriflavine sensitized photopolymer material was used in the development of a panchromatic photopolymer for holographic recording applications [20]. Previously, the characterization of a 60 μm thick transmission grating with high spatial frequencies was reported [27]. In this work, a similar composition is optimized for the range of spatial frequencies polymer layer thickness needed in this application. Figure 7a shows the composition of the layers and Figure 7b, the absorption spectrum for this acriflavine-sensitized photopolymer.

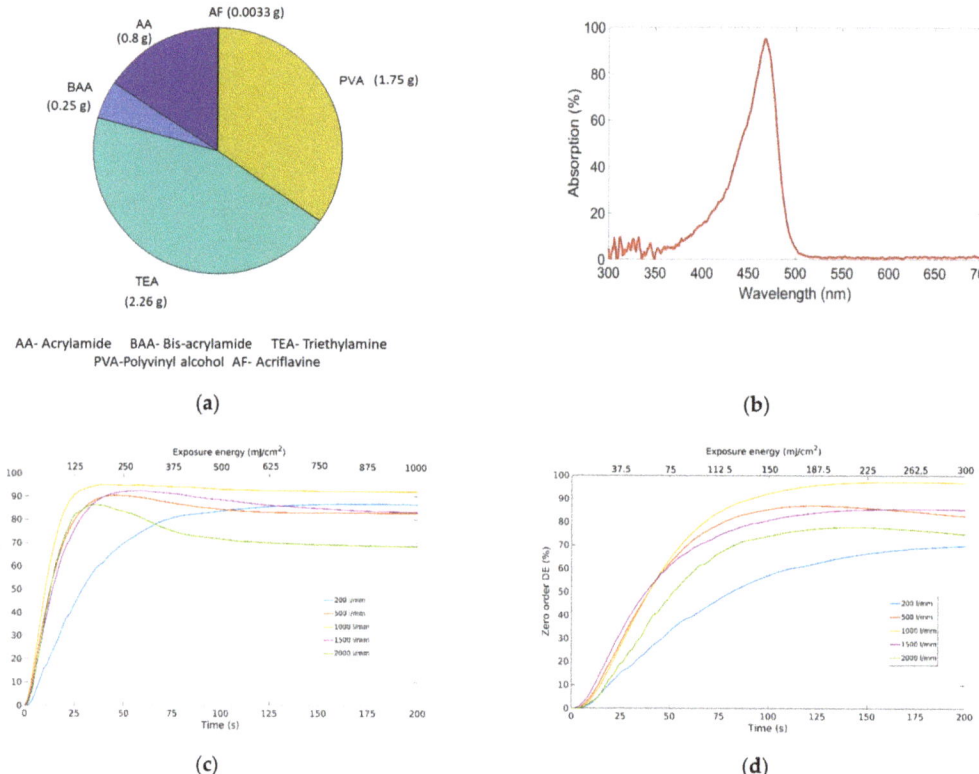

Figure 7. (a) The relative mass of the components of the acriflavine-sensitized dry photopolymer layer. (b) The absorption spectrum of the acriflavine-sensitized photopolymer (composition as outlined in (a)). (c) On-Bragg diffraction efficiency growth curve of 60 μm thick gratings recorded at various spatial frequencies, with recording intensity = 5 mW/cm^2 (d) with recording intensity = 1.5 mW/cm^2.

3.2.1. Optimization for Spatial Frequency

Figure 7 shows the optimization of the recording conditions for a range of spatial frequencies in Figure 7c,d. This study was carried out using 60 micron layers so that detailed information about the grating growth dynamics could readily be obtained using on-Bragg dynamic measurements of the diffraction efficiency during recording. High angular selectivity makes on-Bragg illumination during recording more challenging for 150 micron layers.

Using two different recording intensities, and monitoring on Bragg at 633 nm, the recording dynamics were observed for spatial frequencies from 200 to 2000 lines/mm (equivalent to 5 micron to 0.5 micron fringe spacing). It should be noted that the zero-order transmission was monitored and the raw data from the zero-order transmittance were transformed into the first-order efficiency. In the diffraction efficiency curves the growth dynamics are very similar for all spatial frequencies across the range, except for the two extremes of 200 and 2000 lines/mm. At the higher end, the 2000 lines/mm curves are observed to fall slightly below the mid-range curves. At 200 lines/mm, a significantly slower grating growth is observed, although the diffraction efficiency is similar after 200 s for the higher intensity recording. The photopolymer material composition and the absorption spectrum are also shown.

3.2.2. Optimization for 150 Micron Thick Layers

Real-time monitoring of the diffraction efficiency during grating recording is considerably more challenging at 150 microns, so a stepwise recording with multiple samples is sometimes needed to properly examine the grating growth, especially at higher spatial frequencies (due to the high angular selectivity). In this section, a combination of real time recording and multiple Bragg curves is used to study blue-sensitive layers of 150 micron thickness. Firstly, real-time recording was done for 200 L/mm spatial frequency with higher (5 mW/cm^2) and lower (1.5 mW/cm^2) intensities. The growth curves of first-order efficiencies are shown in Figure 8a.

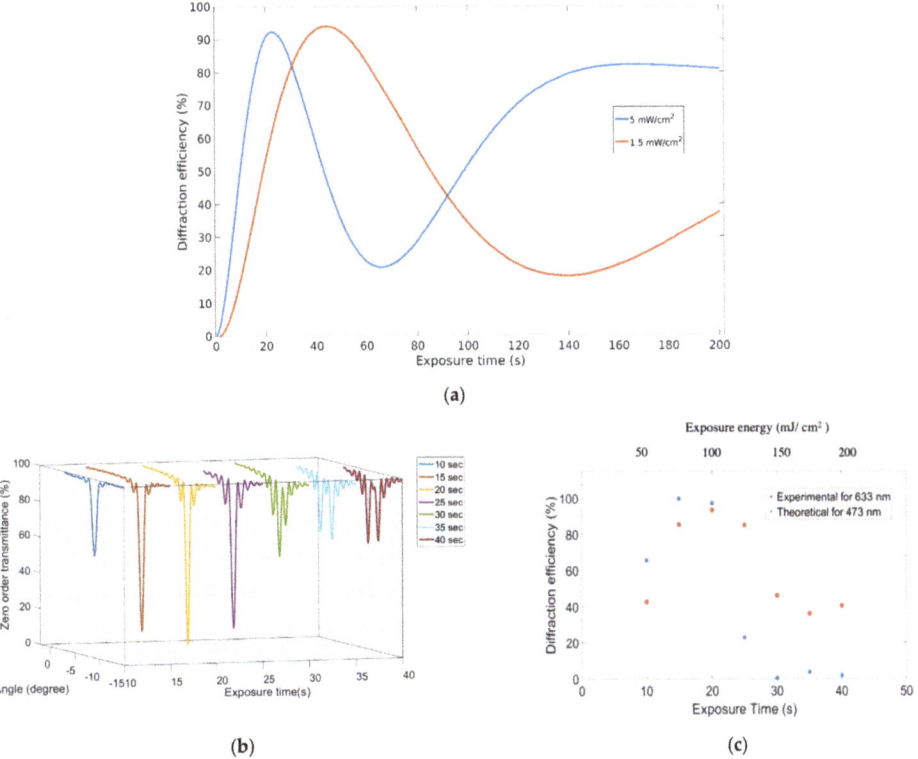

Figure 8. Diffraction efficiency growth for 150 micron thick blue-sensitive layers. (**a**) On-Bragg efficiency growth curve of for gratings recorded with spatial frequency 200 L/mm, (**b**) a series of zero-order Bragg selectivity curves for grating recorded with 500 L/mm, for a range of exposure times, measured at 633 nm. (**c**) On-Bragg diffraction efficiency values obtained from (**b**). The equivalent diffraction efficiency that would be expected at 488 nm is also shown so that the optimal recording time can be identified.

For 500 L/mm, individual gratings were recorded at a constant intensity (5 mW/cm^2) but with different exposure times. The recording time for each gratin g was increased from 10 to 40 s with an interval of 5 s. Figure 8b shows the results obtained in the form of zero-order Bragg selectivity curves, when the individual gratings were probed after recording by a He–Ne laser. Each curve graphs a set of data obtained by rotating the probe beam though a range of angles while monitoring the diffraction efficiency. The curves are arranged in Figure 8b according to the exposure time used to create the grating. They show gratings of increasing strength up to the optimum exposure and then the characteristic indications of overmodulation (growing sidelobes). The minimum value from each zero-order Bragg selectivity was taken to obtain the growth curve, which is shown in Figure 8c, along with theoretical values for the efficiency, which is for the same strength grating at 488 nm (using Equation (3)). Considering the growth of diffraction efficiency for both 200 L/mm and 500 L/mm spatial frequencies, the optimized time to record gratings with 5 mW/cm^2 for λ = 488 nm is estimated ≈16–20 s. As seen in Figure 6, the holographic lens to be recorded covers the range of spatial frequencies from 198 L/mm to 790 L/mm.

3.3. Optimization of the Holographic Recording Conditions for the Fabrication of Individual Lenses at Blue and Green Wavelengths

The next step was to record and test the individual lenses. As lenses consist of many spatial frequencies and slant angles in a single element, it is necessary to determine the optimal recording conditions for each lens, as well as to analyze the overall diffraction behavior to ensure that the expected microstructure is recorded. This makes it possible to (i) confirm experimentally which exposure conditions (from the range identified above) give the best performance across the lens aperture and (ii) to provide performance data of individual lenses as a comparator for the stacked lenses.

3.3.1. Recording and Characterization of Holographic Lens Designed for 488 nm

Individual lenses were recorded at a constant recording intensity with a number of different exposure times. Each lens was then characterized by probing it at three positions across the element through a wide angular range in order to measure the local DE and angular positions of the diffraction peaks (troughs). This allowed for fine tuning of efficiency and uniformity across the lens aperture as well as investigation of the local spatial frequency and slant of the recorded photonic structure. The spatial frequency at each location is indicated by the angle between the two diffraction maxima (minima) in the intensity values of the zero-order wide angular scan, and the slant can be estimated from their offset from 0° at each location. For convenience, this is done at a single wavelength (633 nm).

The recording of these individual lenses was done at a constant recording intensity of 3.6 mW/cm^2 with 17, 27 and 37 s exposure times, thereby achieving total exposure energies of 61, 97 and 130 mJ/cm^2, respectively. After recording, the lenses were probed at three positions across the element (the center of the lens and +/− 3 mm either side). The zero order and first order intensity curves are shown in each case for a range of incident angles in Figure 9. The latter is useful because it is a direct confirmation that the light diffracted from the zero-order curve appears in the first order, validating the assumption that volume gratings/Bragg diffractions as beams other than those of the first-order are of negligible intensity. A number of observations can be made from these results. Firstly, the uniformity of the diffraction efficiency is an important factor in the performance of the collimating device so by looking at the diffraction behavior at three known locations, the recording can be optimized for uniformity as well as the magnitude of the diffraction efficiency. Secondly, the spatial frequency of the grating can be examined by studying the separation of the diffraction peaks/troughs. From Table 2 below, we can see that the separations expected for a 633 nm probe beam for the lens element G1 (designed for 488 nm light) are 13.4°, 18.0° and 22.5° respectively. Taking Figure 9a as an example, we observe separations of 12.1°, 16.8° and 22.8°. These match the expected values to within 1.2°. The differences are probably due to errors in positioning of the probe beam and an averaging effect caused by the probe beam spanning a non-zero range of locations/grating periods. There is also some variation between the three samples.

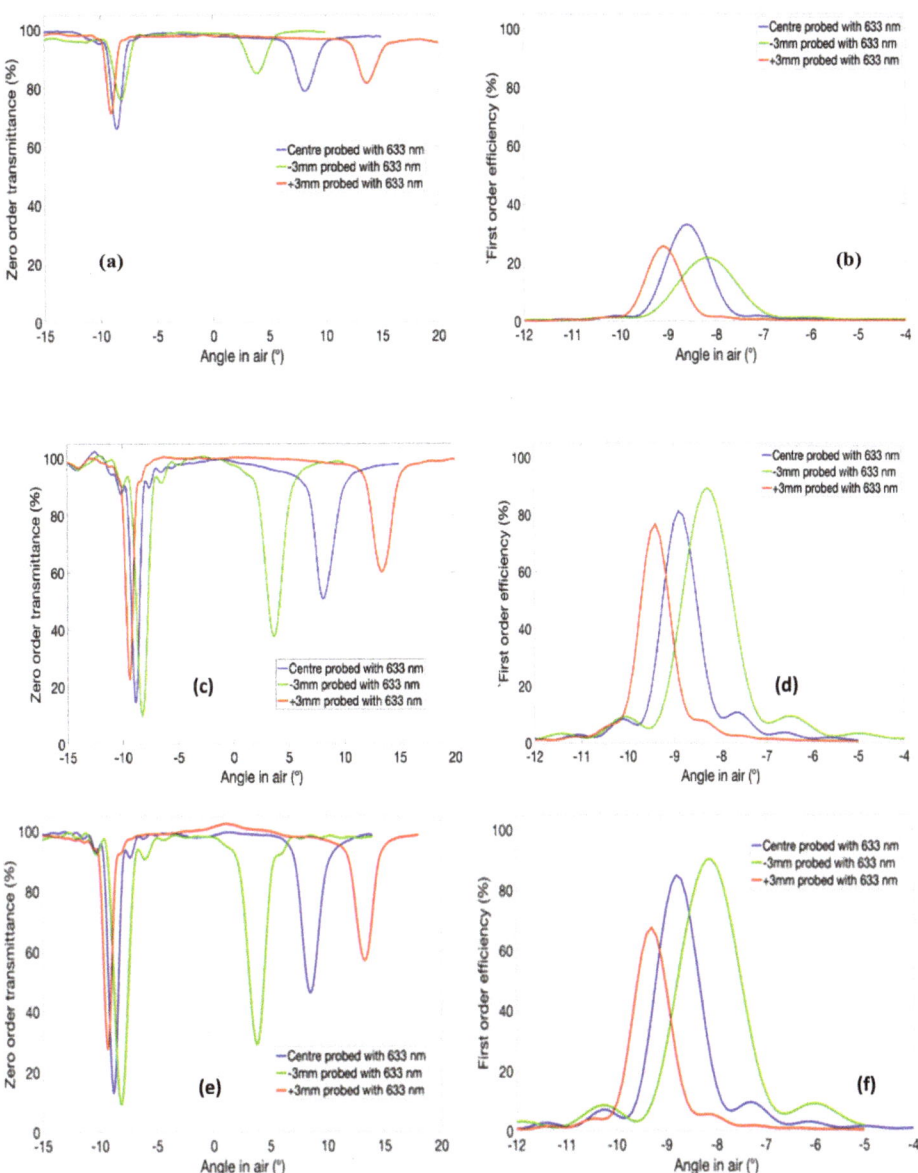

Figure 9. Experimental data for (**a**) zero-order transmission and (**b**) first-order diffraction efficiency at three points across the lens fabricated in acriflavine-sensitized photopolymer layer, using exposure 61 mJ/cm^2, exposure 97 mJ/cm^2 (**c**) and (**d**) exposure 130 mJ/cm^2 (**e**) and (**f**).

Table 2. The measured diffraction angle (angular separation of +1 and −1 diffraction orders) in comparison to the theoretical diffraction angles for the 4 samples at 633 nm wavelength.

Position on Lens (mm)	Measured Diffraction Angle (Degrees)	Measured Diffraction Angle (Degrees)	Measured Diffraction Angle (Degrees)	Measured Diffraction Angle (Degrees)	Theoretical Diffraction Angle for 633 nm (Degrees)	
	Sample 1	Sample 2	Sample 3	Sample 4	G1 Lens for 488 nm	G2 Lens for 514 nm
−3 mm	13.1	13.1	11.2	12.4	13.4	11.9
Center	17.8	18.3	16.9	17.8	18.0	16.0
+3 mm	23.0	23.4	22.1	22.8	22.6	20.0

3.3.2. Recording and Characterization of Holographic Lens Designed for 514 nm

The process described in Section 3.3.1 was repeated for a holographic lens, designed to operate at 514 nm, and recorded in an erythrosine-B-sensitized photopolymer layer. This time, the recording wavelength was 514 nm and recording energies of 98 mJ/cm^2 and 70 mJ/cm^2 were used. From Table 2 below, we can see that the separations expected for a 633 nm probe beam for the lens element G2 (designed for 514 nm light) are 11.9, 16.0 and 20.0. Taking Figure 10a as an example, we observe separations of 10.5°, 15.0° and 20.3°. These match the expected values to within 1.4°. However, the match appears to be poorer in Figure 10c.

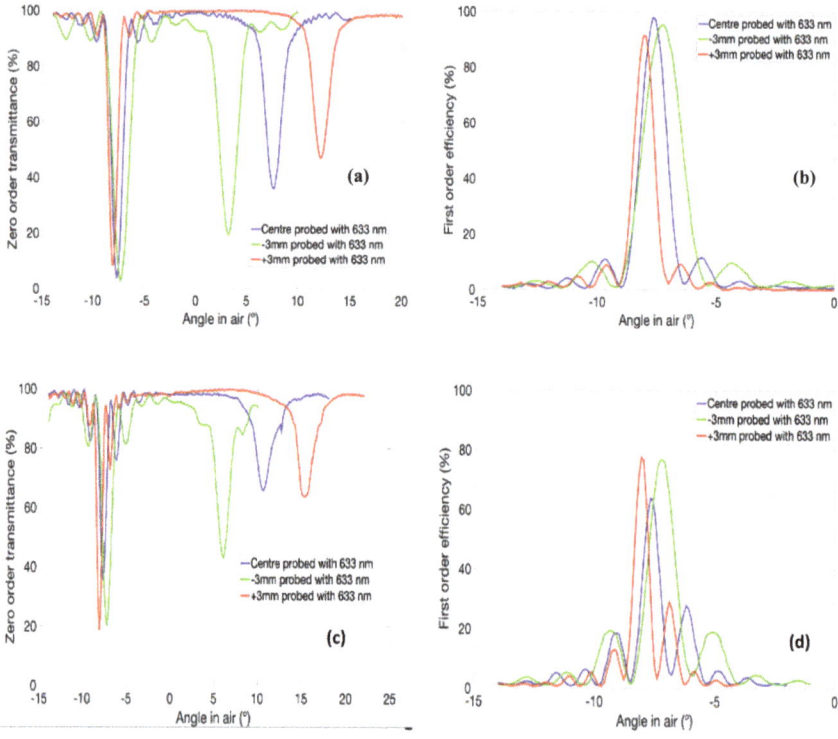

Figure 10. Experimental data for (**a**) the zero-order transmission and (**b**) the first-order diffraction efficiency at three points across the lens fabricated in a 150 μm thick eythrosine-B sensitized photopolymer layer. (Exposure energy = 70 mJ/cm^2) and for (**c**) and (**d**) with exposure energy = 98 mJ/cm^2.

3.4. Two Lenses in Stacked in Dual-Sensitive Photopolymer Layer

For the final device, the dual-sensitized layer was prepared as described above. Two lenses were then recorded sequentially using the same optical set-up (described above) but switching an Ar-Ion lasers lasing wavelength from 514 nm and 488 nm between the recording, using the lasers coupling prism. For each exposure, the intensity and duration followed the optimized conditions found in the previous sections. In the object beam, a conventional cylindrical lens was introduced to create a diverging beam in one axis, which interfered with a collimated reference beam. The resulting interference pattern was recorded in each photopolymer layer. Firstly, a laser (λ = 514 nm) records the lens pattern in the erythrosine-B-sensitized photopolymer layer, which is the second layer in the stack. Secondly, the laser at 488 nm records the lens in the acriflavine-sensitized photopolymer material, which is the first layer in the stack. After recording, the stacked lenses were probed at three positions (center, +3 mm and −3 mm), over a wide range of angles using a 633 nm wavelength laser to obtain the zero-order transmission angular selectivity curve. A set of four stacked lenses were recorded with the same optimized recording conditions (from Section 3.3) and their results are shown in Figure 11.

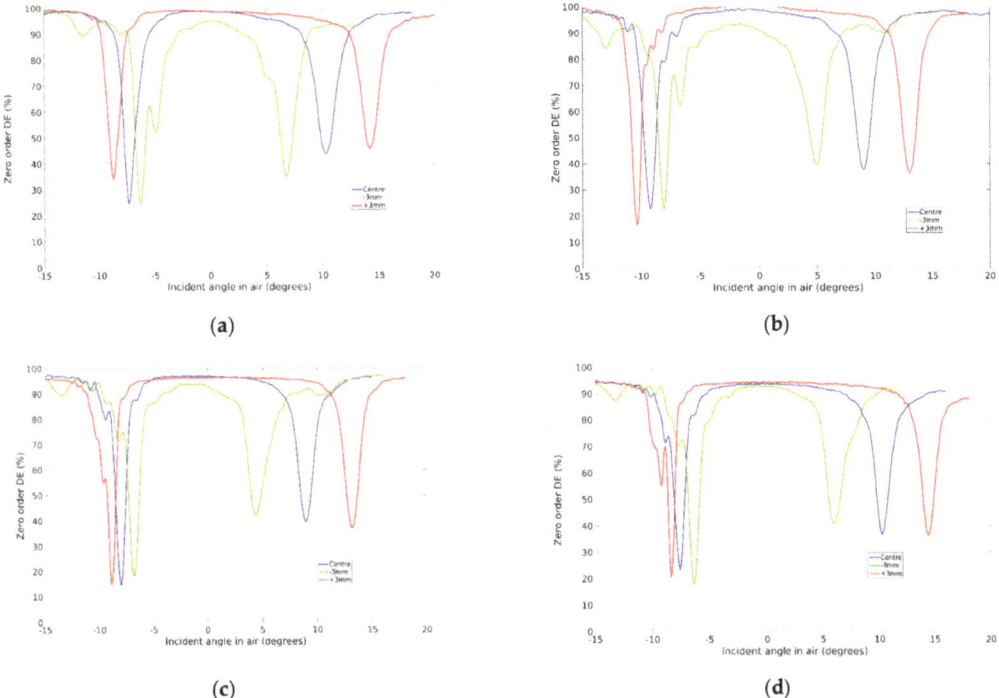

Figure 11. Zero-order transmission angular selectivity curves when the stacked lens is probed by a He–Ne laser at three positions. (**a**), (**b**), (**c**) and (**d**) show the results for samples (1), (2), (3) and (4) respectively.

In Table 2, the angular separation of the incident angles at which maximum diffraction occurs for 633 nm light is tabulated for the three positions in all four samples. A theoretical angular separation is also shown, based on the diffraction angle expected for the recorded spatial frequency at that location. Neither lens is designed for 633 nm light, which is why some secondary peaks were observed (not noted in the table). The peak positions are closer to the positions for the lens designed for blue light, except in the case of sample 2, possibly because this is the lens that the incident beam meets first in the stack.

In order to examine the diffraction angles at the target wavelengths, sample 1 was then tested at the design/recording wavelengths. Using a manual rotation stage, the angular position of the diffraction peaks was obtained, and the angular separation or full diffraction angle was calculated. Table 3 shows the measured diffraction angle (angular separation of +1 and −1 diffraction orders) in comparison to the theoretical diffraction angles for both gratings in the stack while probing at the recording/design wavelengths. In most cases (+3 mm G1 is the exception) there is better than 0.5° agreement between the measured diffraction angle for the stack with the expected diffraction angle for that wavelength.

Table 3. Table showing the measured diffraction angle (angular separation of +1 and −1 diffraction orders) in comparison to the theoretical diffraction angles for both gratings in the stack while probing at the recording/design wavelengths.

Probe Wavelength (nm)	Position on Lens (mm)	Measured Diffraction Angle (Degrees)	Theoretical Diffraction Angle (Degrees)	
			G1 lens designed for blue 488 nm	G2 lens designed for green 514 nm
514.5	Center	13.3	14.1	13.4
	−3 mm	10.0	10.5	10.0
	+3 mm	18.3	17.8	16.9
488	Center	13.0	13.4	12.8
	−3 mm	10.3	10.0	9.5
	+3 mm	17.0	16.9	16.0

The stack was then illuminated with laser light. The input beam was a horizontally diverging beam produced by the cylindrical lens (the same lens that was used in recording). It was arranged that this be incident on the stack at the appropriate angle and the laser was then switched between wavelengths of 488 nm and 514 nm. In both cases, an intense diffracted beam was clearly observed (to the right in Figure 12). While the transmitted portion of the input beam (seen on the left in each figure) was highly diverging, the diffracted output beam was observed to have much lower divergence and to have a circular cross-section. This demonstrates that the stack functions well for both wavelengths. As shown in Figure 12c, the process was repeated with a white light which was collimated and passed through the same cylindrical lens. Again, the stack produced a strong diffracted beam with a circular cross section (to the right in the image). Of course, some dispersion of the colors in the white light source is also evident, especially for the red and violet colors at the extreme ends of the visible spectrum.

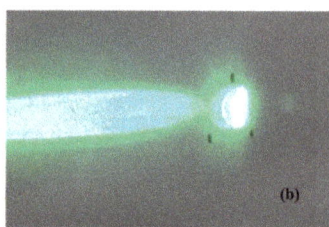
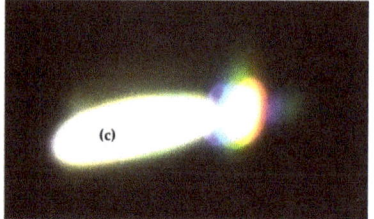

Figure 12. Demonstration of the stacked lens producing a collimated beam when illuminated with a cylindrically expanding laser beam at (a) 488 nm and (b) 514 nm and (c) with a white LED passing through a cylindrical lens.

4. Discussion

The exposure optimization work for blue recording showed the viability of using this blue-sensitized photopolymer for recording low spatial frequency gratings in layers as thick as 150 microns. High diffraction efficiencies were obtained with a range of exposures. This material had not previously been characterized in this range. The recording of the DOE lenses holographically in individual layers showed the challenges of equalizing efficiency across a DOE that contains a range of spatial frequencies, but 80–90% was achieved for green recording and greater than 90% for the blue recordings. With optimization, this could be improved further. It should be noted that the results indicate that two structures were recorded with the microstructure that matched well to the theoretical modeling. However, because the two gratings were designed to have overlapping input and output beams when they function properly, it is not possible to resolve the two individual sets of grating structures. Even at wavelengths far from the design wavelengths, modeling showed that the peaks for G1 and G2 would be less than $0.5°$ apart. In addition, limitations in repeatable positioning of the probe beam on the lens meant that there were relatively large errors in the actual peak positions as well as some averaging over a range of neighboring locations. This will be improved by using a smaller diameter probe beam and devising a system of fiducial marks to properly locate the recording center.

5. Conclusions

In this paper, two photopolymer layers sensitive to two different wavelengths were stacked together and lenses were recorded in each layer. Initially, a theoretical design was developed in order to determine the recording conditions needed to achieve an efficient element. An acriflavine-sensitized photopolymer material was characterized so as to find the recording conditions for the expected range of spatial frequencies. Holographic lenses were then recorded in individual layers at blue and green wavelengths (in the appropriately sensitized photopolymer), which demonstrated uniform and high efficiency at the optimized exposure energies. Dual-wavelength sensitive stacks were then assembled and the sequential recording of stacked lenses was carried out. A set of four stacked lenses were recorded and tested by probing by a He–Ne laser at a wide range of angles for three positions across the element. The angular separation of +1 and −1 order maximum diffraction positions were analyzed. It was found that experimental angular separation matched well with theoretical values for the lenses. It was not possible to resolve the separate diffraction peaks from the blue and green lenses, as they were designed to coincide for blue and green wavelengths; even at 633 nm, the angular separation is expected to be $< 0.5°$. Initial tests performed with cylindrically diverging beams showed strong diffraction into collimated off-axis beams for blue, green and white light. With some optimization, this approach could be used to develop stacked DOEs that can re-direct and collimate light with multiple wavelength ranges for potential applications in illumination, solar collection and displays.

Future work will involve investigating the dual-wavelength layer stacking approach for lenses with different off-axis angles and numerical apertures as well as a more detailed study of the independent photonic structures recorded by this method by deliberately separating the positions of the diffraction peaks.

Author Contributions: S.K. carried out most of the experimental work and developed aspects of the methodology as well as preparing the original draft. He acquired the funding for the work as part of a PhD scholarship. B.R. contributed to the experimental work in the final section. K.M. helped develop the methodology and contributed to draft preparation and data analysis. He is a co-supervisor of this work. K.R. assisted with the modeling and data analysis. I.N. helped develop the concept, methodology and materials. She is a co-supervisor of this work. S.M. helped develop the concept methodology and materials. She is the lead-supervisor of this work. All supervising authors contributed to reviewing and editing. All authors have read and agreed to the published version of the manuscript.

Funding: This research was funded by TU Dublin Fiosraigh PhD Scholarship fund. This research received no funding external to TU Dublin.

Acknowledgments: The authors would like to acknowledge Vincent Toal for his advice and the FOCAS research Institute for facilities and support.

Conflicts of Interest: The authors declare no conflict of interest.

References

1. Kogelnik, H. Coupled Wave Theory for Thick Hologram Gratings. *Bell Syst. Tech. J.* **1969**, *48*, 2909–2947. [CrossRef]
2. Close, D.H. Holographic Optical Elements. *Opt. Eng.* **1975**, *14*, 408–419. [CrossRef]
3. Motogaito, A.; Machida, N.; Morikawa, T.; Manabe, K.; Miyake, H.; Hiramatsu, K. Fabrication of a binary diffractive lens for controlling the luminous intensity distribution of LED light. *Opt. Rev.* **2009**, *16*, 455–457. [CrossRef]
4. Mohammad, N.; Meem, M.; Wan, X.; Menon, R. Full-color, large area, transmissive holograms enabled by multi-level diffractive optics. *Sci. Rep.* **2017**, *7*, 1–6. [CrossRef] [PubMed]
5. Baldry, I.; Bland-Hawthorn, J.; Robertson, J.G. Volume Phase Holographic Gratings: Polarization Properties and Diffraction Efficiency. *Publ. Astron. Soc. Pac.* **2004**, *116*, 403–414. [CrossRef]
6. Bruder, F.K.; Fäcke, T.; Rölle, T. The Chemistry and Physics of Bayfol® HX Film Holographic Photopolymer. *Polymers* **2017**, *9*, 472. [CrossRef]
7. Martin, S.; Akbari, H.; Keshri, S.; Bade, D.; Naydenova, I.; Murphy, K.; Toal, V. Holographically Recorded Low Spatial Frequency Volume Bragg Gratings and Holographic Optical Elements. *Hologr. Mater. Opt. Syst.* **2017**. [CrossRef]
8. Wang, H.; Wang, J.; Liu, H.; Yu, D.; Sun, X.; Zhang, J. Study of effective optical thickness in photopolymer for application. *Opt. Lett.* **2012**, *37*, 2241–2243. [CrossRef]
9. Bianco, G.; Ferrara, M.A.; Borbone, F.; Roviello, A.; Striano, V.; Coppola, G. Photopolymer-based volume holographic optical elements: Design and possible applications. *J. Eur. Opt. Soc. Rapid Publ.* **2015**, *10*, 15057. [CrossRef]
10. Marín-Sáez, J.; Atencia, J.; Chemisana, D.; Collados, M.-V. Characterization of volume holographic optical elements recorded in Bayfol HX photopolymer for solar photovoltaic applications. *Opt. Express* **2016**, *24*, A720–A730. [CrossRef]
11. Orselli, E.; Zanutta, A.; Bianco, A.; Karafolas, N.; Cugny, B.; Sodnik, Z.; Fäcke, T. Photopolymer materials for volume phase holographic optical elements. In Proceedings of the International Conference on Space Optics Vol. 10562—ICSO 2016, Biarritz, France, 18–21 October 2016.
12. Han, J.; Liu, J.; Yao, X.; Wang, Y. Portable waveguide display system with a large field of view by integrating freeform elements and volume holograms. *Opt. Express* **2015**, *23*, 3534–3549. [CrossRef]
13. Khan, M.S.; Rahlves, M.; Lachmayer, R.; Roth, B. Polymer-based diffractive optical elements for rear end automotive applications: Design and fabrication process. *Appl. Opt.* **2018**, *57*, 9106–9113. [CrossRef]
14. Marín-Sáez, J.; Keshri, S.; Naydenova, I.; Murphy, K.; Atencia, J.; Chemisana, D.; Garner, S.; Collados, M.V.; Martin, S. Stacked holographic optical elements for solar concentration. In Proceedings of the Photonics Ireland Conference, Cork, Ireland, 3–5 September 2018.
15. Piao, M.L.; Kwon, K.C.; Kang, H.J.; Lee, K.Y.; Kim, N. Full-color holographic diffuser using time-scheduledmiterative exposure. *Appl. Opt.* **2015**, *54*, 5252–5259. [CrossRef]
16. Shen, Z.; Lan, T.; Wang, L.; Ni, G. Color demultiplexer using angularly multiplexed volume holograms as a receiver optical end for VLC based on RGB white LED. *Opt. Commun.* **2014**, *333*, 139–145. [CrossRef]
17. Avayu, O.; Almeida, E.; Prior, Y.; Ellenbogen, T. Composite functional metasurfaces for multispectral achromatic optics. *Nat. Commun.* **2017**, *8*, 14992. [CrossRef] [PubMed]
18. Hong, K.; Yeom, J.; Jang, C.; Hong, J.; Lee, B. Full-color lens-array holographic optical element for three-dimensional optical see-through augmented reality. *Opt. Lett.* **2014**, *39*, 127–130. [CrossRef]
19. Malallah, R.; Li, H.; Qi, Y.; Cassidy, D.; Muniraj, I.; Al-Attar, N.; Sheridan, J.T. Improving the uniformity of holographic recording using multi-layer photopolymer: Part II Experimental results. *J. Opt. Soc. Am. A* **2019**, *36*, 334–344. [CrossRef] [PubMed]
20. Meka, C.; Jallapuram, R.; Naydenova, I.; Martin, S.; Toal, V. Development of a panchromatic acrylamide-based photopolymer for multicolor reflection holography. *Appl. Opt.* **2010**, *49*, 1400–1405. [CrossRef] [PubMed]
21. Mahamat, A.H.; Narducci, F.A.; Schwiegerling, J. Design and optimization of a volume-phase holographic grating for simultaneous use with red, green, and blue light using unpolarized light. *Appl. Opt.* **2016**, *55*, 1618–1624. [CrossRef]
22. Jauncey, G.E.M. The Scattering of X-Rays and Bragg's Law. *Proc. Natl. Acad. Sci. USA* **1924**, *10*, 57–60. [CrossRef]
23. Moharam, M.G.; Young, L. Criterion for Bragg and Raman-Nath diffraction regimes. *Appl. Opt.* **1978**, *17*, 1757–1759. [CrossRef] [PubMed]
24. Keshri, S.; Murphy, K.; Toal, V.; Naydenova, I.; Martin, S. Development of a photopolymer holographic lens for collimation of light from a green light-emitting diode. *Appl. Opt.* **2018**, *57*, E163–E172. [CrossRef] [PubMed]
25. Akbari, H.; Naydenova, I.; Martin, S. Using acrylamide-based photopolymers for fabrication of holographic optical elements in solar energy applications. *Appl. Opt.* **2014**, *53*, 1343–1353. [CrossRef] [PubMed]
26. Pramitha, V. A New Metal Ion Doped Panchromatic Photopolymer for Holographic Applications. Ph.D. Thesis, Cochin University of Science and Technology, Cochin, India, November 2011. Available online: dyuthi.cusat.ac.in/purl/2357 (accessed on 1 May 2021).
27. Meka, C. Development of Acrylamide Based Photopolymer for Full Colour Display Holography. Ph.D. Thesis, Technological University Dublin, Dublin, Ireland, 2010. [CrossRef]

Article

Double Spectral Electromagnetically Induced Transparency Based on Double-Bar Dielectric Grating and Its Sensor Application

Guofeng Li [1,2], Junbo Yang [2], Zhaojian Zhang [2], Kui Wen [2], Yuyu Tao [1], Yunxin Han [2,*] and Zhenrong Zhang [1,*]

1. Guangxi Key Laboratory of Multimedia Communications and Network Technology, School of Computer, Electronics and Information, Guangxi University, Nanning 530004, China; liguofeng24@163.com (G.L.); yuyu.tao@foxmail.com (Y.T.)
2. Center of Material Science, National University of Defense Technology, Changsha 410073, China; yangjunbo@nudt.edu.cn (J.Y.); 376824388@alumni.sjtu.edu.cn (Z.Z.); kuiwen93@hotmail.com (K.W.)
* Correspondence: hanyx15@nudt.edu.cn (Y.H.); zzr76@gxu.edu.cn (Z.Z.)

Received: 26 March 2020; Accepted: 24 April 2020; Published: 27 April 2020

Abstract: The realization of the electromagnetically induced transparency (EIT) effect based on guided-mode resonance (GMR) has attracted a lot of attention. However, achieving the multispectral EIT effect in this way has not been studied. Here, we numerically realize a double EIT-ike effect with extremely high Q factors based on a GMR system with the double-bar dielectric grating structure, and the Q factors can reach 35,104 and 24,423, respectively. Moreover, the resonance wavelengths of the two EIT peaks can be flexibly controlled by changing the corresponding structural parameters. The figure of merit (FOM) of the dual-mode refractive index sensor based on this system can reach 571.88 and 587.42, respectively. Our work provides a novel method to achieve double EIT-like effects, which can be applied to the dual mode sensor, dual channel slow light and so on.

Keywords: guided-mode resonance; electromagnetically induced transparency; high quality factor; double spectral; refractive index sensor

1. Introduction

In the past few decades, the electromagnetically induced transparency (EIT) effect caused by quantum destructive interference between two different excitation paths in a three-level atomic system has attracted extensive and in-depth research [1,2]. The EIT effect is that under the effect of strong resonance coupling light, the opaque medium becomes transparent for weak probe light and is accompanied by extraordinarily steep dispersion. In other words, the resonance probe light can propagate through the medium without being absorbed [3]. EIT has potential applications in many fields, such as sensing [4–6], slow light [5,7], nonlinear optics [8] and cavity quantum electrodynamics [9]. However, the realization of the EIT effect in atomic systems requires extremely strict experimental conditions, which limits its practical application. Therefore, researchers have turned to classical optical systems to achieve EIT-like effects. G. Shevts and J.S Wertele described the EIT analog in a classic plasma [10], and more of an EIT-like effect was subsequently observed in the plasmonic metamaterial [11,12]. The EIT-like effect has also been successfully achieved in other optical systems, such as waveguide cavity structures [13,14] and photonic crystals [15]. Sun-Goo Lee's group realized the EIT-like effect based on the guided-mode resonance (GMR) effect for the first time in 2015 [16], the system consists of two planar dielectric waveguides and a subwavelength grating and a narrow transparent window appears in the transmission dip when high Q and low Q resonant waveguide modes are coupled.

In recent years, the realization of the EIT-like effect based on GMR has attracted more and more interest of researchers. Sun-Goo Lee's group successfully realizes the polarization-independent EIT-like effect in two photonic systems, which are both composed of two planar dielectric waveguides and a two-dimensional photonic crystal [17]. Sun Y and Chen H et al. reported a planar metamaterial based on GMR achieve the EIT-like effect with a Q value exceeding 7000 [18]. Han Y and Yang et al. reported that a GMR system with two subwavelength silicon grating waveguide layers achieves the EIT-like effect with an ultra-high Q factor of 288,892 [19]. However, as is known to us, the realization of multispectral EIT effects by the subwavelength grating structure based on GMR has not been studied according to available works.

In this work, we numerically simulate the double spectral EIT-like effect in a coupled GMR system that consists of two silicon grating waveguide layers (GWLs) on a SiO_2 substrate. In addition, both GWLs contain double-bar dielectric gratings with unequal lengths in a cycle. When the distance between the two GWLs meets the phase matching condition, the GMR system works as a Fabry–Perot (F–P) cavity and GWLs as the reflection boundary of the cavity thus two EIT peaks appear in the near infrared band due to the top GMR mode and the bottom GMR mode are intercoupled. The effects of two grating widths and the separation between the two gratings are also studied. After the structural parameters optimization, two EIT peaks present ultra-narrow FWHMs of 10^{-2} nm and ultra-high Q factors of 10^4. In terms of the sensor, this system has ordinary sensitivity and high figure of merit (FOM) with compact structure and a simple manufacturing process. This double spectral EIT-like effect may not only be applied to the sensor, but also to slow light and nonlinear optics.

2. Structure and Simulation

The 3D finite-difference time-domain (FDTD) method was used to study the phenomenon of this optical system. During the simulation with FDTD, the mesh accuracy was set to $\lambda/34$ to ensure the convergence of the simulated results. The schematic of the GMR system with a double-bar dielectric grating structure is shown in Figure 1a. In terms of boundary conditions, X and Y were set to periodic, and Z was the perfectly matched layer (PML). The mesh accuracy in GWL_1 and GWL_2 were $\Delta x = \Delta y = 5$ nm and $\Delta z = 10$ nm, respectively. The inset of Figure 1b shows the real part of the refractive index value for SiO_2 [20] and Si [21] as a function of the wavelength. The background index of the system was $n_s = 1$.

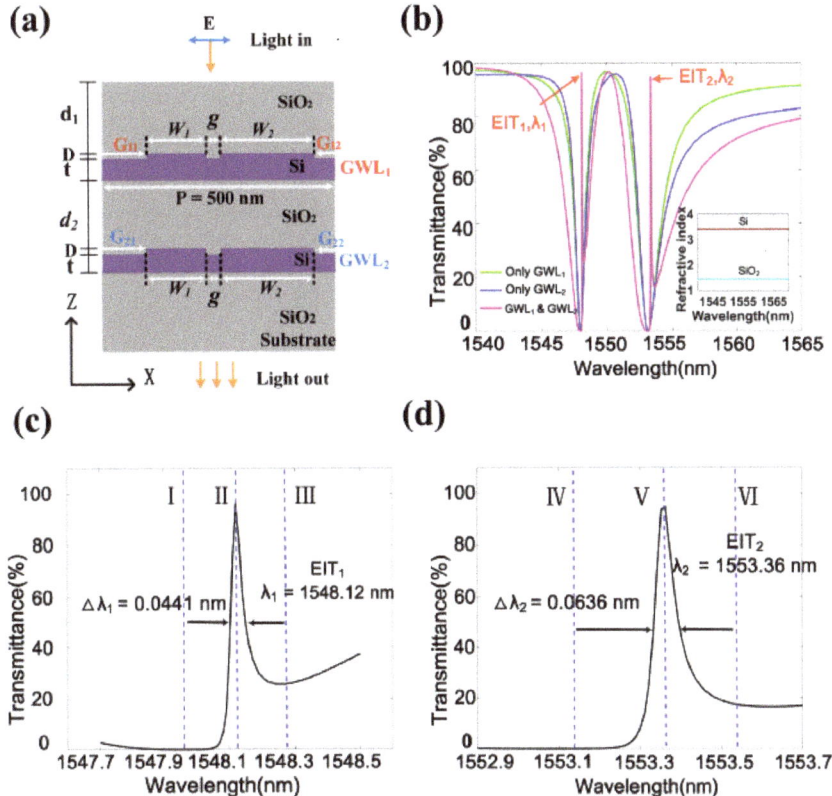

Figure 1. (a) Schematic of the guided-mode resonance (GMR) system with double-bar dielectric grating structure and geometrical parameters: d_1 = 1885 nm, D = 55 nm, t = 435 nm, d_2 = 2137 nm, W_1 = 100 nm, W_2 = 230 nm, g = 30 nm and P = 500 nm. (b) Transmittance spectra of the GMR system with only grating waveguide layer $(GWL)_1$ (GWL_2) and both GWLs. The inset shows the real part of the refractive index for SiO_2 and Si as a function of the wavelength. (c) Magnified view of the transmission features of electromagnetically induced transparency $(EIT)_1$. (d) Magnified view of the transmission features of EIT_2.

Light polarized in the X direction (TM polarization) incidents vertically above the system in the near-infrared spectral band. The period P of grating was 500 nm. A SiO_2 layer with a thickness of 1885 nm was designed on the top of the GMR structure in order to reduce the reflection of light. The distance (d_2) between the two GWLs was designed to be 2137 nm. Gratings G_{11} and G_{12} had the same depths of D = 55 nm, and different widths of W_1 = 100 nm and W_2 = 230 nm, respectively. The separation between two Si gratings was g = 30 nm. The Si waveguide thickness was t = 435 nm. Meanwhile, structural parameters of GWL_1 and GWL_2 were consistent.

3. Results and Discussion

We first calculated the transmission spectrum of the structure with only single GWL_1 (GWL_2), as the green solid line (blue solid line) shown in Figure 1b. It is clear that the two different transmission dips appeared at the wavelengths of 1547.97 nm and 1553.08 nm. GWL_1 and GWL_2 had the same structure, but their optical properties were easily affected by the surrounding medium, thus the transmission spectra were slightly different. The SiO_2 cover layer had an influence on the resonance frequency. Adding a SiO$_2$ cover layer can reduce the impact of the surrounding medium on its optical characteristics to reach the same resonance frequency [18].

When GWL$_1$ and GWL$_2$ are both present, and their distance (d_2) met the phase matching conditions, the top GMR mode (in GWL$_1$) and the bottom GMR mode (in GWL$_2$) were intercoupled, Therefore, two different sharp EIT resonances appear in the two resonance dips respectively due to destructive interference [12,19,22]. The transmission spectrum is shown by the solid red line in Figure 1b. We refer to the two EIT peaks as EIT$_1$ and EIT$_2$. The transmission characteristics of their magnified views are shown in Figure 1c,d, respectively. The resonance wavelengths of EIT$_1$ (λ_1) and EIT$_2$ (λ_2) were 1548.12 nm and 1553.36 nm, respectively, and their corresponding full width at half maximum (FWHM) was 0.0441 nm ($\Delta\lambda_1$ in Figure 1c) and 0.0636 nm ($\Delta\lambda_2$ in Figure 1d). The Q factor was an important parameter of the resonant cavity, which is defined as follows:

$$Q = \frac{\lambda}{\text{FWHM}} \quad (1)$$

There, their corresponding Q factors reached 35,104 and 24,423, respectively. Obviously, our work has two higher Q factors than another double EIT-like effects work, which has two Q factors of 950 and 216 [6].

The analogy between our system and the atomic EIT system can help us understand the physics of the double EIT-like effect [23,24]. A double-Λ five-level model of the double EIT-like effect in our system is illustrated in Figure 2a. In the system, |1⟩ represents the ground state, the field of the bottom GMR mode corresponds to the probability amplitudes of atoms in the metastable states |2⟩ and |3⟩, the field of the top GMR mode corresponds to the probability amplitudes of atoms in the excited states |4⟩ and |5⟩ [19,23,25]. The control field refers to the coupling between the top GMR mode and the bottom GMR mode, and the probe field refers to the input of the top GMR mode. We observed that two sharp EIT-like windows appeared in the probe area due to the introduction of the control field. The energy level of |5⟩ — |1⟩ was higher than |4⟩ — |1⟩, hence EIT$_1$ corresponded to the destructive interference between two different transition pathways |1⟩ → |5⟩ and |1⟩ → |5⟩ → |3⟩ → |5⟩, EIT$_2$ corresponded to the destructive interference between two different transition pathways |1⟩ → |4⟩ and |1⟩ → |4⟩ → |2⟩ → |4⟩.

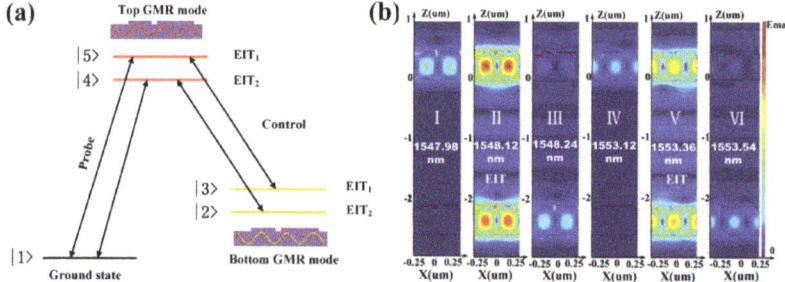

Figure 2. (a) Double-Λ five-level model of the double EIT-like effect in this GMR system. (b) The electric field distribution diagrams of the GMR system correspond to the wavelengths indicated by the blue dash lines in Figure 1c,d, and I at the off-resonant wavelength of EIT$_1$ 1547.98nm, II at the EIT$_1$—resonant wavelength of 1548.12 nm, III at the off-resonant wavelength of EIT$_1$ 1548.24 nm, IV at the off-resonant wavelength of EIT$_2$ 1553.12 nm, V at the EIT$_2$—resonant wavelength of 1553.36 nm and VI at the off-resonant wavelength of EIT$_2$ 1553.54 nm.

In order to help us understand the origin of the double EIT-like effect in this GMR system, electric field distribution diagrams of the GMR system near two EIT peaks (such as the transmission spectra of Figure 1c,d are given in Figure 2b. Near the EIT$_1$, electric field distribution at the off—resonant wavelength of EIT$_1$ 1547.98 nm (the blue dash I in Figure 1c) and 1548.24 nm (the blue dash III in

Figure 1c) was very easily observed and excitation mainly occurred in GWL$_1$ (I in Figure 2b) or GWL2 (III in Figure 2b), which was consistent with the characteristics of GMR. The corresponding transmittance was very low since the GMR mode was easily coupled to electromagnetic waves in free space. Near the EIT$_2$, the electric field distribution at the off—resonant wavelength of EIT$_2$ 1553.12 nm (the blue dash IV in Figure 1d) and 1553.54 nm (the blue dash VI in Figure 1d) corresponded to IV and VI in Figure 2b. The principles were the same as near EIT$_1$.

At the EIT$_1$—resonant wavelength of λ_1 = 1548.12 nm (the blue dash II in Figure 1b) and the EIT$_2$—resonant wavelength of λ_2 = 1553.36 nm (the blue dash V in Figure 1d), the electric field distribution diagrams were II and V Figure 2b, respectively. Obviously, very strong oscillations occurred simultaneously in GWL$_1$ and GWL$_2$. When the incident light was at two EIT wavelengths, the light was reflected back and forth between GWL$_1$ and GWL$_2$, and electromagnetic energy was coupled into the two GWLs, so strong oscillations were excited through coupling. Once the top GMR mode and the bottom GMR mode are intercoupled, and this system works like as an F–P cavity [19,26,27], therefore, two very narrow transmission peaks could be obtained at the wavelengths of 1548.12 nm and 1553.36 nm in the transmission spectrum, where transmissions reached 96.23% and 94.48%, respectively.

The transmission spectra of the coupled GMR system with different d_2 are shown in Figure 3, and other parameters are the same as Figure 1. In the range of d_2 from 1970 to 2220 nm, with the increase of d_2, the two resonance wavelengths (λ_1 and λ_2) were both red-shifted. When d_2 was increased from 2020 (in Figure 3b) to 2120 nm (in Figure 3d), λ_1 was increased from 1547.67 to 1548.06 nm and λ_2 was increased from 1552.51 to 1553.22 nm. This verified that the GMR system worked like an F–P cavity and GWLs as the reflective boundary of the cavity when d_2 met the phase matching condition. Obviously, for the bottom GMR mode in GWL$_2$, a displacement of about one hundred nanometers could achieve a favorable coupling with the top GMR mode in GWL$_1$, two EIT peaks appeared in the two transmission dips when the F–P cavity was introduced. Generally, a phase matching condition corresponds to a resonance wavelength when the distance and refractive index are determined [19]. Here, it is worth mentioning that when the determined d_2 satisfies the phase matching condition, two EIT peaks with different resonant wavelengths can be generated simultaneously because the refractive index of Si is different at different wavelengths.

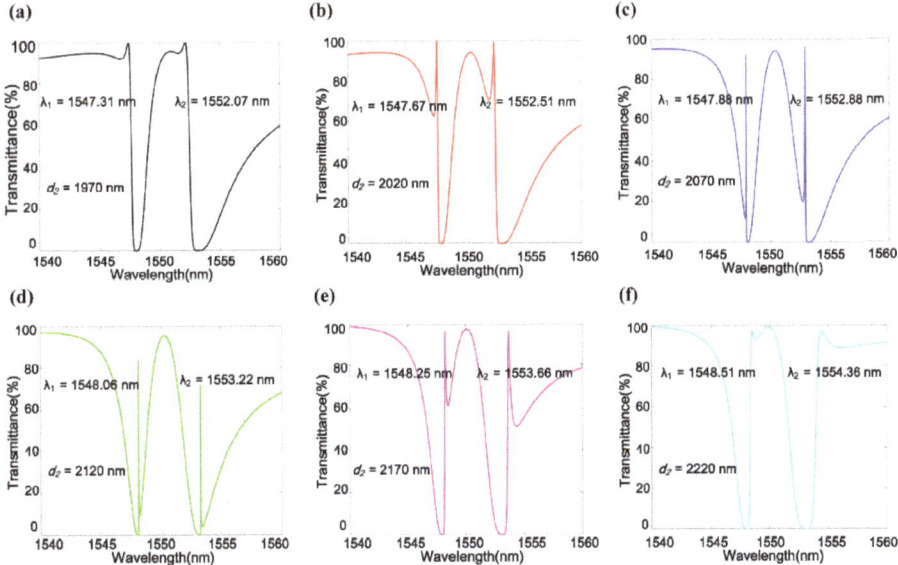

Figure 3. Transmittance spectra of the coupled GMR systems with different d_2. (**a**) Transmission spectrum with d_2 = 1970 nm. (**b**)Transmission spectrum with d_2 = 2020 nm. (**c**) Transmission spectrum

with d_2 = 2070 nm. (**d**) Transmission spectrum with d_2 = 2120 nm. (**e**) Transmission spectrum with d_2 = 2170 nm. (**f**) Transmission spectrum with d_2 = 2220 nm.

The transmission spectra and resonance wavelengths under different W_1 and W_2 are shown in Figure 4, and other parameters are the same as Figure 1. Distinctly, with the increase of W_1 from 85 to 145 nm or the increases of W_2 from 180 to 240 nm, both λ_1 and λ_2 appeared red shifted. For a clearer observation, we show two resonance wavelengths under different W_1 and W_2 in Figure 4c,d. When W_1 increased from 85 to 145 nm, λ_1 moved from 1547.33 to 1549.93 nm and λ_2 moved from 1552.16 to 1558.13 nm. When W_2 increased from 180 to 240 nm, λ_1 moved from 1545.44 to 1549.2 nm and λ_2 moved from 1549.41 to 1554.54 nm. Thus we could easily control the two resonance wavelengths by adjusting the parameters of W_1 or W_2. The reason for the red shift of the two resonance wavelengths is that as W_1 or W_2 increase, the fill factor increases, which changes the average refractive index of the grating layer [28].

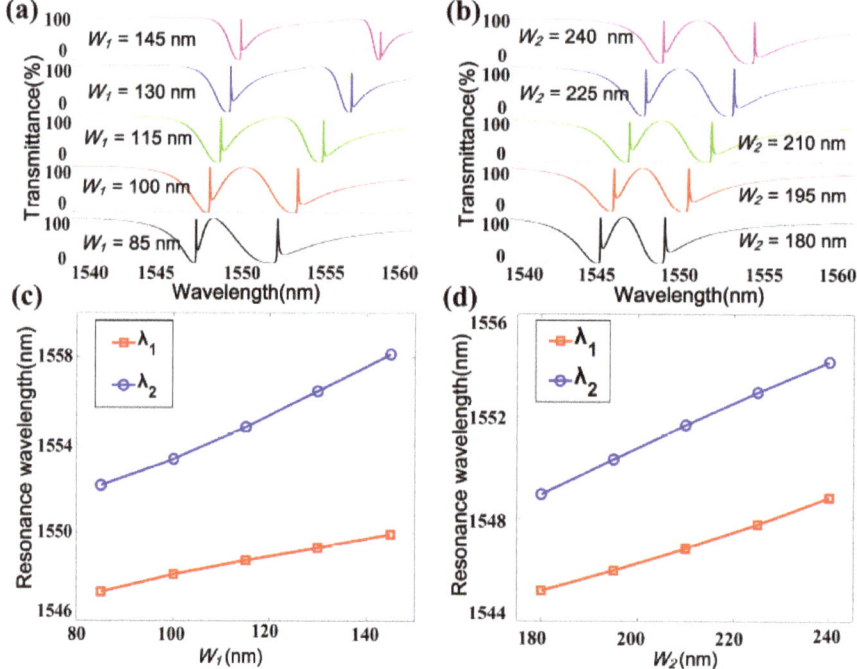

Figure 4. (**a**) The transmission spectra under different W_1 and W_1 is equal to 85 nm, 100 nm, 115 nm, 130 nm and 145 nm, respectively. (**b**) The transmission spectra under different W_2 and W_2 is equal to 180 nm, 195 nm, 210 nm, 225 nm and 240 nm, respectively. (**c**) Resonance wavelengths of two EIT peaks at different W_1. (**d**) Resonance wavelengths of two EIT peaks at different W_2.

The gap between the gratings G_{11} and G_{12} (gratings G_{21} and G_{22}) is also an important parameter, being defined as "g". The transmission spectrums under different g are shown in Figure 5 and other parameters are the same as Figure 1. Obviously, as g varied from 10 to 50 nm in steps of 10 nm, the resonance wavelength of EIT_1 (λ_1) appeared blue shifted, while the resonance wavelength of EIT_2 (λ_2) appeared red shifted. Two resonance wavelengths and two resonance wavelengths difference ($\lambda_2-\lambda_1$) when g changed from 10 to 50 nm are shown in Figure 5f. As g increased, λ_1 moved from 1549.93 to 1547.19 nm and λ_2 moved from 1552.55 to 1553.93 nm (as shown on the left vertical axis in Figure 5f), and "λ_2-λ_1" increased from 2.62 to 6.74 nm (as shown in the right vertical axis in Figure 5f). The resonance wavelength of the two EIT peaks could also be flexibly adjusted by adjusting the parameter g.

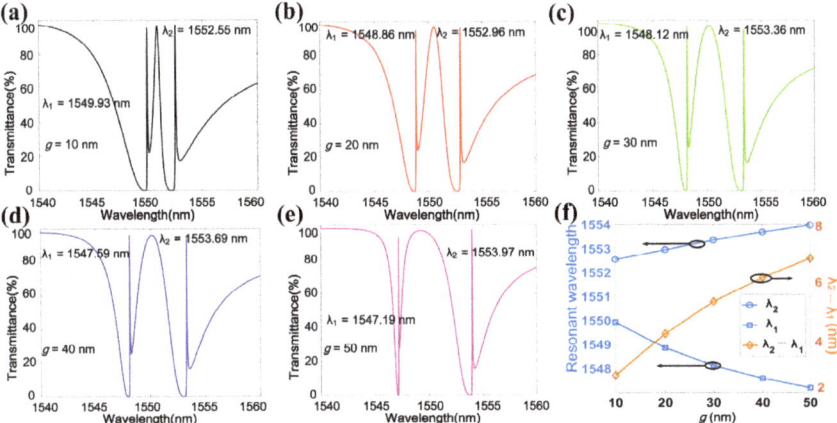

Figure 5. The transmission spectra under different g, other parameters are the same as Figure 1. (**a**) g = 10 nm. (**b**) Transmission spectrum with g = 20 nm. (**c**) Transmission spectrum with g = 30 nm. (**d**) Transmission spectrum with g = 40 nm. (**e**) Transmission spectrum with g = 50 nm. (**f**) Two resonance wavelengths and two resonance wavelengths difference (λ_2-λ_1) under different g.

Sensing Performance

Considering the resonance wavelength difference between the two EIT peaks and the transmittance of the two EIT peaks (not shown here), we chose a structure with W_1 = 100 nm, W_2 = 230 nm and g = 30 nm to study the corresponding sensing performance. At the same time, the SiO_2 between GWL_1 and GWL_2 changed to the dielectric sample. In order to make the refractive index change of the dielectric more sensitive, d_2 was set to 3740 nm. The performance of the sensor was evaluated by two factors, sensitivity (S) and FOM [29]:

$$S = \frac{\Delta\lambda}{\Delta n} \quad (2)$$

$$FOM = \frac{S}{FWHM} \quad (3)$$

Here, S refers to the resonance wavelength shift caused by the change in the refractive index unit of the dielectric, and FOM represents the optical resolution of the sensor.

Figure 6 shows the transmission spectrum with a different refractive index of the dielectric, the insert shows the resonance wavelengths of two EIT peaks at a different refractive index of the dielectric. It is obvious that the resonance wavelengths of the two modes had a linear relationship with the refractive index of the dielectric. The changes in EIT_1 and EIT_2 with the refractive index of the dielectric are referred to as mode 1 and mode 2, respectively. Therefore, the average FWHM of the mode 1 and mode 2 from the refractive index of the dielectric from 1.440 to 1.452 was 0.051 nm and 0.061 nm. The S of mode 1 and mode 2 were 29.166 nm/RIU and 35.833 nm/RIU, respectively, so FOM of mode 1 was 571.88, and FOM of mode 2 was 587.42. Compared with the other sensors, this sensor had ordinary sensitivity due to the narrow line width of two EIT peaks [30]. However, it had higher FOM than other previous sensors [5,6,31,32], as shown in Table 1. So, this sensor had a super high optical resolution.

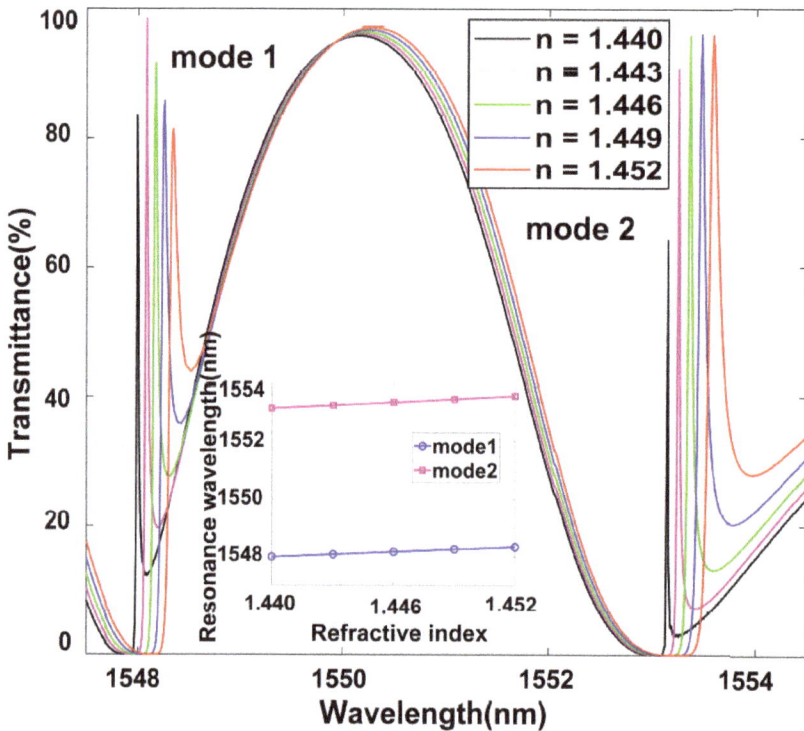

Figure 6. Transmission spectrum of the refractive index of the dielectric between the GWL$_1$ and the GWL$_2$ changed from 1.440 to 1.452. The inset shows the resonance wavelengths of the EIT$_1$ peak (mode 1) and the EIT$_2$ peak (mode 2) when the refractive index of the dielectric changes from 1.440 to 1.452.

Table 1. Figure of merit (FOM) compared with previous sensors.

Reference	FOM$_1$	FOM$_2$
[5]	42.00	
[6]	330.00	281.00
[31]	39.00	
[32]	20.00	
this work	571.88	587.42

4. Conclusions

In summary, we proposed a GMR system with a double-bar dielectric grating structure to achieve a double spectral EIT-like effect. The F–P cavity was introduced due to the distance (d_2) between the two GWLs satisfying the phase matching condition, and two EIT peaks with ultra-narrow FWHM and ultra-high Q factors could be obtained. The influences of the width and separation of the double-bar gratings on two EIT peaks were also investigated. Meanwhile, the performance of the ultra-high FOM sensors based on this system was also studied. This work provided a way to achieve double spectral EIT-like effect, which has potential applications in the dual mode sensor, dual channel slow light and so on.

Author Contributions: G.L. completed structural design, simulation calculation and writing of articles. Y.H. provided the research ideas and paper revision of this article. J.Y. and Z.Z. (Zhenrong Zhang) determined the research direction, revised papers and provided fund support. Z.Z. (Zhaojian Zhang), K.W. and Y.T. contributed to the software setting and analysis of simulation results. All authors have read and agreed to the published version of the manuscript.

Funding: This research received no external funding.

Acknowledgments: This work was supported by the National Natural Science Foundation of China under Grant 61661004, Guangxi Science Foundation (2017GXNSFAA198227), National Natural Science Foundation of China (60907003, 61805278), the China Postdoctoral Science Foundation (2018M633704), the Foundation of NUDT (JC13-02-13, ZK17-03-01), the Hunan Provincial Natural Science Foundation of China (13JJ3001), and the Program for New Century Excellent Talents in University (NCET-12-0142).

Conflicts of Interest: The authors declare that they have no competing interests.

References

1. Kocharovskaya, A.; Khanin, Y.I. Population trapping and coherent bleaching of a three-level medium by a periodic train of ultrashort pulses. *Zh. Eksp. Teor. Fiz* **1986**, *90*, 1610–1618.
2. Harris, S.E. Electromagnetically induced transparency. In *Quantum Electronics and Laser Science Conference (p. QTuB1)*; Optical Society of America: Washington, DC, USA, 1997.
3. Hau, L.V.; Harris, S.E.; Dutton, Z.; Behroozi, C.H. Light speed reduction to 17 metres per second in an ultracold atomic gas. *Nature* **1999**, *397*, 594–598. [CrossRef]
4. Liu, N.; Weiss, T.; Mesch, M.; Langguth, L.; Eigenthaler, U.; Hirscher, M.; Giessen, H. Planar metamaterial analogue of electromagnetically induced transparency for plasmonic sensing. *Nano Lett.* **2010**, *10*, 1103–1107. [CrossRef] [PubMed]
5. Wei, Z.; Li, X.; Zhong, N.; Tan, X.; Zhang, X.; Liu, H.; Liang, R. Analogue electromagnetically induced transparency based on low-loss metamaterial and its application in nanosensor and slow-light device. *Plasmonics* **2017**, *12*, 641–647. [CrossRef]
6. Li, W.; Su, Y.; Zhai, X.; Shang, X.; Xia, S.; Wang, L. High-Q Multiple Fano Resonances Sensor in Single Dark Mode Metamaterial Waveguide Structure. *IEEE Photonics Technol. Lett.* **2018**, *30*, 2068–2071. [CrossRef]
7. Huang, Y.; Min, C.; Veronis, G. Subwavelength slow-light waveguides based on a plasmonic analogue of electromagnetically induced transparency. *Appl. Phys. Lett.* **2011**, *99*, 143117. [CrossRef]
8. Ilchenko, V.S.; Savchenkov, A.A.; Matsko, A.B.; Maleki, L. Nonlinear optics and crystalline whispering gallery mode cavities. *Phys. Rev. Lett.* **2004**, *92*, 043903. [CrossRef]
9. Xiao, Y.F.; Özdemir, Ş.K.; Gaddam, V.; Dong, C.H.; Imoto, N.; Yang, L. Quantum nondemolition measurement of photon number via optical Kerr effect in an ultra-high-Q microtoroid cavity. *Opt. Express* **2008**, *16*, 21462–21475. [CrossRef]
10. Shvets, G.; Wurtele, J.S. Transparency of magnetized plasma at the cyclotron frequency. *Phys. Rev. Lett.* **2002**, *89*, 115003. [CrossRef]
11. Strelniker, Y.M.; Bergman, D.J. Transmittance and transparency of subwavelength-perforated conducting films in the presence of a magnetic field. *Phys. Rev. B* **2008**, *77*, 205113. [CrossRef]
12. Liu, N.; Langguth, L.; Weiss, T.; Kästel, J.; Fleischhauer, M.; Pfau, T.; Giessen, H. Plasmonic analogue of electromagnetically induced transparency at the Drude damping limit. *Nat. Mater.* **2009**, *8*, 758–762. [CrossRef] [PubMed]
13. Yanik, M.F.; Suh, W.; Wang, Z.; Fan, S. Stopping light in a waveguide with an all-optical analog of electromagnetically induced transparency. *Phys. Rev. Lett.* **2004**, *93*, 233903. [CrossRef]
14. Xu, Q.; Sandhu, S.; Povinelli, M.L.; Shakya, J.; Fan, S.; Lipson, M. Experimental realization of an on-chip all-optical analogue to electromagnetically induced transparency. *Phys. Rev. Lett.* **2006**, *96*, 123901. [CrossRef] [PubMed]
15. Yang, X.; Yu, M.; Kwong, D.L.; Wong, C.W. All-optical analog to electromagnetically induced transparency in multiple coupled photonic crystal cavities. *Phys. Rev. Lett.* **2009**, *102*, 173902. [CrossRef] [PubMed]
16. Lee, S.G.; Jung, S.Y.; Kim, H.S.; Lee, S.; Park, J.M. Electromagnetically induced transparency based on guided-mode resonances. *Opt. Lett.* **2015**, *40*, 4241–4244. [CrossRef] [PubMed]
17. Lee, S.G.; Kim, S.H.; Kim, K.J.; Kee, C.S. Polarization-independent electromagnetically induced transparency-like transmission in coupled guided-mode resonance structures. *Appl. Phys. Lett.* **2017**, *110*, 111106. [CrossRef]

18. Sun, Y.; Chen, H.; Li, X.; Hong, Z. Electromagnetically induced transparency in planar metamaterials based on guided mode resonance. *Opt. Commun.* **2017**, *392*, 142–146. [CrossRef]
19. Han, Y.; Yang, J.; He, X.; Huang, J.; Zhang, J.; Chen, D.; Zhang, Z. High quality factor electromagnetically induced transparency-like effect in coupled guided-mode resonant systems. *Opt. Express* **2019**, *27*, 7712–7718. [CrossRef]
20. Malitson, I.H. Interspecimen comparison of the refractive index of fused silica. *Josa* **1965**, *55*, 1205–1209. [CrossRef]
21. Palik, E.D. (Ed.) *Handbook of Optical Constants of Solids*; Academic Press: New York, NY, USA, 1998.
22. Papasimakis, N.; Fedotov, V.A.; Zheludev, N.I.; Prosvirnin, S.L. Metamaterial analog of electromagnetically induced transparency. *Phys. Rev. Lett.* **2008**, *101*, 253903. [CrossRef]
23. Fleischhauer, M.; Imamoglu, A.; Marangos, J.P. Electromagnetically induced transparency: Optics in coherent media. *Rev. Mod. Phys.* **2005**, *77*, 633. [CrossRef]
24. Zhang, S.; Genov, D.A.; Wang, Y.; Liu, M.; Zhang, X. Plasmon-induced transparency in metamaterials. *Phys. Rev. Lett.* **2008**, *101*, 047401. [CrossRef]
25. Liu, Y.C.; Li, B.B.; Xiao, Y.F. Electromagnetically induced transparency in optical microcavities. *Nanophotonics* **2017**, *6*, 789–811. [CrossRef]
26. Kekatpure, R.D.; Barnard, E.S.; Cai, W.; Brongersma, M.L. Phase-coupled plasmon-induced transparency. *Phys. Rev. Lett.* **2010**, *104*, 243902. [CrossRef]
27. Zhang, J.; Liu, W.; Yuan, X.; Qin, S. Electromagnetically induced transparency-like optical responses in all-dielectric metamaterials. *J. Opt.* **2014**, *16*, 125102. [CrossRef]
28. Qi, W. *Study on the Mechanism and Characteristics of Guided-Mode Resonance Subwavelength Device*; University of Shanghai for Science and Technology: Shanghai, China, 2012; pp. 49–51.
29. Zhang, Z.; Yang, J.; Xu, H.; Xu, S.; Han, Y.; He, X.; Chen, D. Hybridization-induced resonances with high quality factor in a plasmonic concentric ring-disk nanocavity. *arXiv* **2019**, arXiv:1904.09437.
30. Shakoor, A.; Grande, M.; Grant, J.; Cumming, D.R. One-dimensional silicon nitride grating refractive index sensor suitable for integration with CMOS detectors. *IEEE Photonics J.* **2017**, *9*, 1–11. [CrossRef]
31. Liu, G.D.; Zhai, X.; Wang, L.L.; Wang, B.X.; Lin, Q.; Shang, X.J. Actively tunable Fano resonance based on a T-shaped graphene nanodimer. *Plasmonics* **2016**, *11*, 381–387. [CrossRef]
32. Zhang, S.; Bao, K.; Halas, N.J.; Xu, H.; Nordlander, P. Substrate-induced Fano resonances of a plasmonic nanocube: A route to increased-sensitivity localized surface plasmon resonance sensors revealed. *Nano Lett.* **2011**, *11*, 1657–1663. [CrossRef]

 © 2020 by the authors. Licensee MDPI, Basel, Switzerland. This article is an open access article distributed under the terms and conditions of the Creative Commons Attribution (CC BY) license (http://creativecommons.org/licenses/by/4.0/).

Article

Design and Modelling of a Novel Integrated Photonic Device for Nano-Scale Magnetic Memory Reading

Figen Ece Demirer [1,*,†], Chris van den Bomen [1,†], Reinoud Lavrijsen [1], Jos J. G. M. van der Tol [2] and Bert Koopmans [1]

1. Department of Applied Physics, Eindhoven University of Technology, 5612 AZ Eindhoven, The Netherlands; chrisvdbomen@gmail.com (C.v.d.B.); r.lavrijsen@tue.nl (R.L.); b.koopmans@tue.nl (B.K.)
2. Department of Electrical Engineering, Eindhoven University of Technology, 5612 AZ Eindhoven, The Netherlands; J.J.G.M.v.d.Tol@tue.nl
* Correspondence: ecedmrr1@gmail.com
† These authors contributed equally to this work.

Received: 30 October 2020; Accepted: 15 November 2020; Published: 21 November 2020

Featured Application: On-chip optical reading of magnetic memory processed as ultrathin magnetic claddings on photonic waveguides.

Abstract: Design and simulations of an integrated photonic device that can optically detect the magnetization direction of its ultra-thin (~12 nm) metal cladding, thus 'reading' the stored magnetic memory, are presented. The device is an unbalanced Mach Zehnder Interferometer (MZI) based on InP Membrane on Silicon (IMOS) platform. The MZI consists of a ferromagnetic thin-film cladding and a delay line in one branch, and a polarization converter in the other. It quantitatively measures the non-reciprocal phase shift caused by the Magneto-Optic Kerr Effect in the guided mode which depends on the memory bit's magnetization direction. The current design is an analytical tool for research exploration of all-optical magnetic memory reading. It has been shown that the device is able to read a nanoscale memory bit (400 × 50 × 12 nm) by using a Kerr rotation as small as 0.2°, in the presence of a noise ~10 dB in terms of signal-to-noise ratio. The device is shown to tolerate performance reductions that can arise during the fabrication.

Keywords: integrated photonics; InP; magneto-optic; MZI; mode conversion; PMA; multi-layered thin film; magnetic memory; MOKE; Fourier transformation

1. Introduction

In the modern world, exponentially increasing generation of data and its handling require novel technologies that perform faster and more energy efficiently. To answer this need, optical components are being used in combination with electronic circuitry to improve the speed and bandwidth of data communication and telecommunication. For example, optical interconnections that were once a conceptual design suggestion [1] are currently being used in commercial products replacing slow and heat-dissipating electrical signal communication channels [2,3]. Researchers continue to demonstrate the superior performance circuitry achieved through the integration of photonics into electronics [4–7]. Yet, these advances require back-and-forth signal conversion between optical and electrical domains, which happens to be the new bottleneck in data communication and processing. Addressing this problem requires establishing novel functionalities in photonic devices that will enable a seamless conversion. Furthermore, (integrated) photonics is lacking a simple and fast non-volatile memory function. A huge potential is anticipated for future devices that enable direct inter-conversion of data between the photonic and magnetic (memory) domain without any intermediate electronics steps, cutting down on time and energy costs. This study works with existing non-volatile magnetic memory material technology used in electronics: multilayered ferromagnetic thin-film layers. When the multilayered magnetic material is used as memory material, writing bits into the magnetic memory could be facilitated by recent advances in so-called all-optical switching of magnetization [8,9]. Reading out magnetic bits back into the photonic domain could be achieved via a nonreciprocal magneto-optical process [10–12], while dynamic, on-the-fly reading of magnetic bits could be facilitated by racetrack memory concept [9,13]. In a racetrack memory, magnetic domains (memory bits) move while the material that carries the magnetic domains remain stationary [14,15]. Previously, domain wall velocities up to 1000 ms^{-1} were demonstrated [16,17]. It is in this spirit that our paper focuses on the functionality of on-chip optical reading of magnetic memory processed as ultrathin magnetic claddings on photonic waveguides. To our best knowledge, this is the first study which explores the possibility of on-chip, all-optical magnetic memory reading functionality.

State-of-the-art non-volatile magnetic memory such as spin-transfer torque magnetic random-access memory (STT-MRAM) relies on ferromagnetic multilayered ultrathin films with perpendicular magnetic anisotropy (PMA), in which the magnetization vector is perpendicular to the film plane [18]. Such PMA films turn out to be essential for the advanced schemes used to electrically control the magnetic memory elements but are also known for their relatively large magneto-optical efficiency. A simple layer stack that hosts all relevant physical mechanisms is Ta(4)/Pt(2)/Co(1)/Pt(2)/Co(1)/Pt(2) where the numbers in parenthesis are thickness in nm. Bringing this memory component to the proximity of light confined in a waveguide in a photonic device setting gives rise to magneto-optic interactions, specifically the Magneto-optic Kerr Effect (MOKE). MOKE causes a change in the polarization state of light (Kerr rotation and ellipticity), which changes sign when the magnetization direction of the memory component is flipped [11,12]. In a photonic waveguide context, this gives rise to partial mode conversion between TE and TM modes, which potentially enables reading of the memory bit. However, the MOKE signal is intrinsically small in amplitude, a typical Kerr rotation is around 0.05° for films with an in-plane magnetization in free-space optics [19,20]. In order to increase the efficiency of the mode conversion, we propose the use of PMA magnetic claddings, which have not been seriously addressed yet in a photonic perspective. Such claddings with a perpendicular magnetic orientation are expected to display larger amplitude magneto-optical effects, yet still small quantitatively. This calls for developing novel approaches to amplify the magneto-optical effects while showing the importance of on-chip analytical tools to explore the fundamental mode conversion properties of photonic waveguides with PMA claddings.

To assess the feasibility of using MOKE for on-chip all-optical magnetic memory reading functionality, as well as using it as an analytical tool to quantitatively measure magnetization-induced mode conversion, we investigated specially designed photonic devices whose waveguides are cladded with ultra-thin (12 nm), nano-scale (50 × 400 nm) PMA magnetic memory bits, of the composition mentioned before. By using mathematical models of the designed photonic devices, whose building block performance parameters are chosen according to the InP Membrane on Silicon (IMOS) platform [21], the accuracy of the memory-bit read-out, optical loss and tolerance to noise are tested. It has been shown that the device is able to read a nanoscale memory bit (400 × 50 × 12 nm) by using a Kerr rotation as small as 0.2°, in the presence of a ∼10 dB noise in terms of signal-to-noise ratio (SNR). This paper is structured in the following way. In Section 2 materials and methods are given. Device designs, magneto-optic simulation, mathematical modelling and data analysis topics are covered. In Section 3 the results obtained via the mathematical model are presented for devices with varying degrees of performance parameters. A data analysis technique using Fourier transformation is presented. Lastly, in Section 4, the conclusions are given.

2. Materials and Methods

In this section, materials and geometries of the parts that contribute to the overall device are explained. In addition, the device concept, optical simulation and mathematical modelling methods are explained in the subsections.

The material which stores the magnetic information (memory bit) is a multi-layered ferromagnetic metal thin-film structure, whose stack order is given in the previous section. These multi-layers display PMA, where the magnetization vector is perpendicular to the film plane [22]. PMA is highlighted due to its relatively large magneto-optical efficiency [23]. The multi-layers are placed on top of the waveguides as the top cladding. The rest of the photonic device is fabricated on InP membranes since the devices are based on the IMOS platform [21]. The waveguides have a cross-section of 300 × 400 nm (height and width) and the multi-layered top claddings have the dimensions of 400 × 50 × 12 nm (width, length and height).

2.1. Optical Simulation and Device Concept

Before describing the optical simulation method to quantify the MOKE in waveguides, a brief overview is given on MOKE and its impact on the light confined in waveguides. Following this, the devcie concept is introduced.

MOKE is a type of magneto-optic interaction that takes place when the light reflects from a magnetized material. In polar configuration, the effect causes a change in the light's polarization state which is quantified by Kerr rotation and ellipticity (in angles). Typically, in the literature, MOKE is reported for single reflections. Comparing a single reflection case with our work, more interaction, thus a larger MOKE are expected in waveguides with magnetized top claddings. To our best knowledge, there is no prior work that quantifies the Kerr rotation in a waveguide setting. Therefore, finite-difference time-domain (FDTD) simulations [24] of the waveguides with top-claddings are conducted to estimate the MOKE in the guided modes. In the simulation, multi-layer cladding material is defined by using the magneto-optic constant obtained from the literature [25]. It is seen that the Kerr effect causes conversion between TE and TM modes in the waveguide, comparable to the polarization rotation in free-space optics. The resulting Kerr rotation (θ), ellipticity (ϕ) and optical loss ($Loss_{clad.}$) values obtained for a single memory bit are listed in Table 2. These values are used as inputs for the mathematical model explained in Section 2.2.

The device design is done by considering the key enabler of the magnetic memory reading functionality: a change in the sign of the Kerr rotation upon flipping of the magnetization direction of the memory bit (memory bit "1" and "0"). Assuming the confined light is initially in TE mode, the Kerr rotation $\pm\theta$ ($\theta \ll 1$) leads to an emergent TM mode whose field amplitude is proportional to θ for bit 1 and $-\theta$ for bit 0. Therefore, devices which can probe the phase of the emergent TM mode are explored. Mach-Zehnder Interferometers (MZI) are chosen due their ability to convert the phase difference (between the interfering branches) into intensity difference. Balanced and unbalanced MZI are considered as two candidates for the final design. An unbalanced MZI, which has a defined path length difference between the two branches is chosen due to the noise related issues that cannot be addressed in a balanced MZI. This is further elaborated when the presented results are discussed in Section 3. Since at an initial stage, a device is designed for research and exploration purposes, on-chip light source or detector are not considered. To couple an off-chip laser source and an off-chip detector to the device, mode-selective grating couplers are added to the design.

The device design is shown in Figure 1. In this device, TE mode-selective grating coupler is used to couple the light in. Later, a multi-mode interferometer (MMI) is used to split the light equally into two branches. On the upper branch, the TE mode is converted into TM via the polarization converter. The propagation continued (in TM mode) and a delay line is crossed. On the lower branch, the memory bit (magnetic cladding section) caused the TE mode to partially convert into TM mode due to Kerr rotation (θ). The light from the two branches are merged via another MMI. After interference took place, the resulting intensity is picked up via a TM-selective grating coupler.

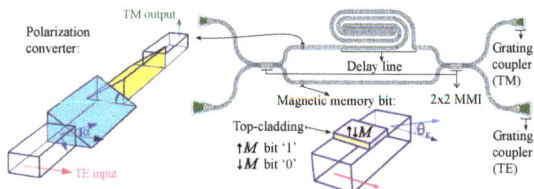

Figure 1. An unbalanced MZI. TE and TM mode selective grating couplers are used to couple the light in and out. The polarization converter is taken from [26].

2.2. Mathematical Modelling and Fourier Transformation

A mathematical model is built in order to simulate the output light intensity vs. light wavelength for the designed devices. The model is given input parameters that are based on IMOS building block performances [21] and FDTD magneto-optical simulations [24] (see Section 2.1). An overview of the model parameters and their brief descriptions are given in Table 1. Additionally, reduced-performance devices with and without noise are simulated with the model to compare the magnetic memory reading capabilities of the devices. These parameters—some standard for all devices and some changing according to the performance levels—are summarized in Tables 2 and 3, respectively.

Using the mathematical model, equations which determine the electric field (E-field) components of TE and TM modes in branches 1 and 2, are obtained. For simplicity, coefficients addressing the losses of mode propagations, grating couplers and magnetic cladding are combined into the terms B_n. For description of other parameters please refer to Table 1.

$$\begin{aligned} E_{TE,1} &= B_1 \cos(\alpha) \, e^{-i \frac{2\pi}{\lambda} n_{TE} L_1} \\ E_{TM,1} &= B_2 \sin(\alpha) \, e^{-i \frac{2\pi}{\lambda} n_{TM} (L_1 - x_{PC})} \\ E_{TE,2} &= B_3 \cos(\theta) e^{-i \frac{2\pi}{\lambda} n_{TE} L_2} \\ E_{TM,2} &= B_4 \sin(\theta) e^{-i \left(\frac{2\pi}{\lambda} n_{TM} (L_2 - x_{\text{clad}}) + \phi \right)}. \end{aligned} \quad (1)$$

Table 1. Overview of model parameters.

Parameters	Definitions
λ	Wavelength scanned (nm)
$Loss_{wg}$	Waveguide propagation loss (assumed to be the same for TE and TM)
n_{TE}	Effective index of TE mode
n_{TM}	Effective index of TM mode
L_1	Length of the upper branch (μm)
L_2	Length of the lower branch (μm)
x_{PC}	Distance between polarization converter and left-hand side MMI splitter (μm)
$x_{\text{clad.}}$	Distance between memory bit (cladding) and left-hand side MMI splitter (μm)
$Loss_{\text{clad.}}$	Loss due to memory bit (cladding)
$Loss_{GC_{TE \to TE}}$	Loss of TE-selective grating coupler for TE mode (dB)
θ	Kerr rotation
ϕ	Kerr ellipticity
α	Angle of mode tilt induced by polarization converter (degree)
$Loss_{GC_{TM \to TM}}$	Loss of TM-selective grating coupler for TM mode (dB)
$Ext_{GC_{TM \to TE}}$	Extinction ratio of TM-selective grating coupler for TE mode (dB)
Noise	Addition of a Gaussian distribution of random noise to the intensity

Table 2. Showing generic parameters that are valid for all devices.

Parameter	Value
λ	1465–1495 nm
$Loss_{wg}$	3 dB/cm
n_{TE}	2.012
n_{TM}	1.809

Parameter	Value
L_1	1386 µm
L_2	462 µm
x_{PC}	200 µm
x_{clad}	100 µm

Parameter	Value
$Loss_{clad.}$	0.13 dB/50 nm
$Loss_{GC_{TE \to TE}}$	1.5 dB
θ	$\pm 0.2°$
ϕ	$\pm 0.2°$

Table 3. Showing parameters that are dependent on the device performance.

Parameter	Standard Device	Reduced Performance	Noise + Reduced Performance
α	90°	45°	45°
$Loss_{GC_{TM \to TM}}$	1.5 dB	7 dB	7 dB
$Ext_{GC_{TM \to TE}}$	50 dB	28 dB	28 dB
Noise	none	none	10.7 dB (SNR)

It is important to recall that the interference takes place between the modes whose E-fields lay in parallel planes and the output light intensity (I) from devices can be calculated via $I = \frac{|E|^2}{2 Z_0}$, where $|E|$ is the total E-field amplitude and Z_0 is the impedance of the vacuum. The presented equations for E-field amplitudes reveal that a wavelength sweep of the input light will result in oscillations in intensity. Recall that the information regarding the magnetization direction of the cladding (memory bit type) can be retrieved from the sign of the Kerr rotation and ellipticity (θ, ϕ). As seen from the equations above, when TE mode input light is used, information of the memory bit type is visible only in the phase of the TM mode output light. For an output light intensity vs. wavelength plot that is obtained upon interference of both TE and TM modes, two oscillation frequencies, ν_{TE} and ν_{TM} that correspond to these modes are observed.

$$\nu_{TE} = \frac{n_{TEg}(L_1 - L_2)}{\lambda^2},$$
$$\nu_{TM} = \frac{n_{TMg}(L_1 - L_2 + x_{clad.} - x_{PC})}{\lambda^2}. \quad (2)$$

n_{TEg} and n_{TMg} in Equation (2) indicate group indices of the respective modes. A Fourier transformation can be applied to the resulting output light intensity vs. wavelength data to separate the TM mode contribution. Thanks to this technique, the amplitude and phase of the TM mode component can be found. In order to separate the TE and TM mode contributions, non-overlapping peaks in the Fourier transform is required. Therefore, at the design stage, it is vital to choose $x_{clad.}$ and x_{PC} parameters (see Table 1) accordingly.

3. Results and Discussion

In order to demonstrate the magnetic memory reading capabilities of our devices, the mathematical model described in Section 2.2 was used. As explained in Section 2.1, the chosen devices were unbalanced interferometers that contain built-in ferromagnetic memory components as their top claddings. The model predicted the output light intensity vs. wavelength plots of the devices with opposing memory bits (bit '1' and '0'). Later these plots were analyzed by the Fourier transformation technique to determine the memory bit type, thus realize 'reading' of the magnetic information. Recall that since the magneto-optic interaction which enables the determination of the memory bit type is only extractable from the phase of the TM mode (when TE mode is used as input), Fourier technique greatly reduced the noise and enhanced the sensitivity.

In Figure 2, the left column plots present output light intensity vs. wavelength data. Note that plots depict the intensity after a windowing function is applied. The right column plots show the Fourier transformation of the left column in blue color and the phase difference between two memory bit states for each oscillatory components in red color. Figure 2a,c,e represent the standard, reduced-performance and noisy reduced-performance devices, respectively. The standard device shown in Figure 2a demonstrate a clear 180° phase shift between the two signals which correspond to the opposite memory states. The Fourier transformation in Figure 2b (right side y-axis) show a single peak which correspond to the TM mode (see Equation (2)). The fact that there is only TM mode is thanks to the well-performing TM-selective out-couplers in the standard devices that have a negligible out-coupling of the TE mode. As expected, the phase difference plot in Figure 2b (left side y-axis) indicate 180° difference at the region which correspond to TM peak. Note that the plots depicting phase difference between two memory states convey meaningful information only at the locations where a correspondent Fourier peak is present. To stress this aspect visually in the graph, the points corresponding to a peak are shown in black, whereas the rest is left grey. In Figure 2c, the 'reduced-performance device' is seen. This device has only 45° conversion at the polarization converter and the TE-mode couples out from the TM-selective out-coupler (see Table 3). Due to coupling out of the TE mode that does not carry information on the memory bit's state, it impossible to observe a 180° phase shift in the intensity vs. wavelength plot upon a change in the memory bit type. As expected, Figure 2d reveals two Fourier peaks that correspond to TE and TM modes. As seen from the peak intensities, despite the use of TM-selective out-couplers, the TE mode dominates. Undeterred by the TE mode dominance, the phase difference plot in Figure 2e indicates a phase of 180° at the position corresponding to the TM peak. The phase shift corresponding to the TE mode reads 0°. Testing the device design further by addition of a noise that described in Section 2.2, Figure 2e,f are obtained. The 'noisy and reduced-performance device' demonstrates that, even though the intensity vs. wavelength plot is dominated by noise and mixed modes, it is still possible to determine the magnetic memory type via the Fourier transform technique.

Referring back to Section 2.1 and clarifying the reason for the choice of an unbalanced MZI design over a balanced one, as seen in Figure 2a, if the device is performing at a fixed wavelength, the change in the light intensity upon changing the memory bit type corresponds to only 0.3% of the total light intensity. This observation indicates that the magnetic memory reading functionality of the device can be obstructed by the noise when operating at a single wavelength. Sweeping of a range of wavelengths accompanied by the Fourier transformation method are the key concepts for eliminating sensitivity to noise and increasing memory reading accuracy. Since the wavelength sweep technique is not successful without the specific frequency oscillations that the added delay line provides, an unbalanced MZI is preferred over a balanced one.

Note that for an ideal device depicted in Figure 2a, the difference in light intensity between the two memory states is proportional to the strength of the Kerr rotation. Therefore, if a calibration by using a material with known Kerr rotation and optical loss is done, very small Kerr rotations can be measured quantitatively by using the same design.

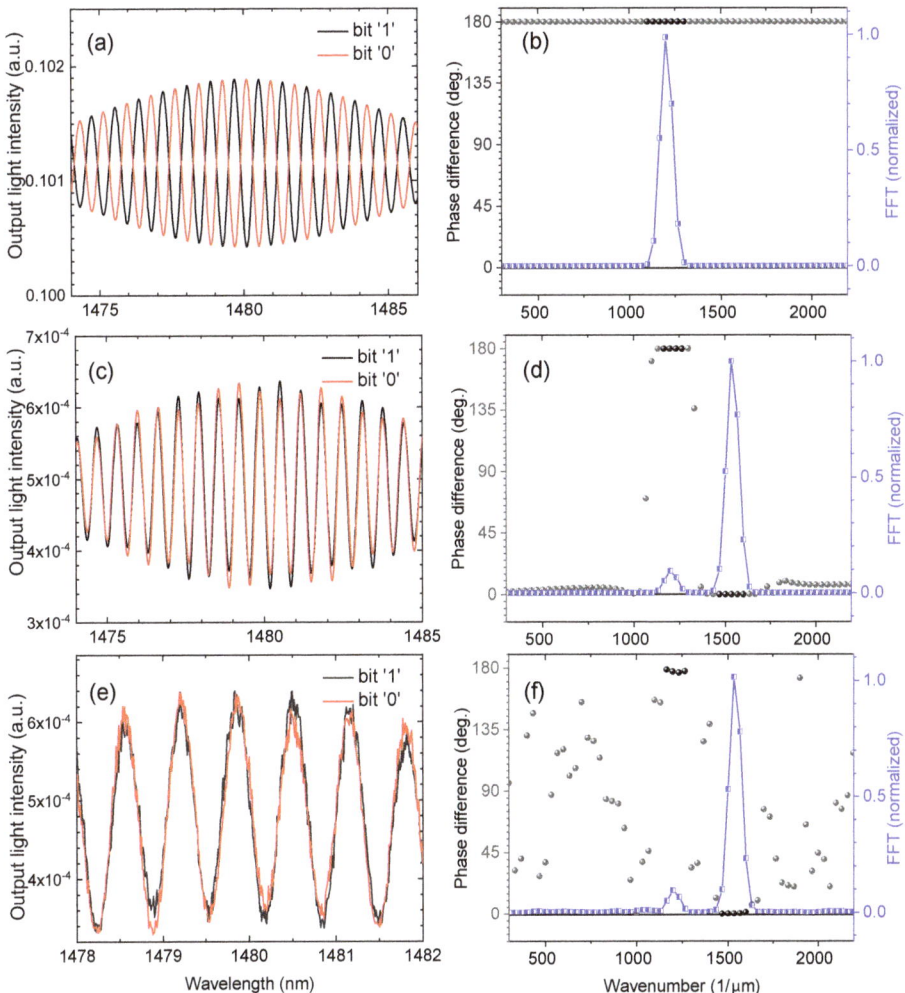

Figure 2. (**a,c,e**) Output light intensity vs. wavelength plots for standard, reduced-performance, and noisy reduced-performance devices (see Tables 2 and 3), The light intensities are shown in arbitrary units and is normalized assuming initial intensity (I_0) is 1. (**b,d,f**) In blue, Fourier transformations of the intensity vs. wavelength plots are shown. The normalization is done assuming the highest intensity Fourier peak has amplitude 1. In black, the phase differences between the memory bit "1" and "0" are shown for each wavenumber. The data-points which correspond to a Fourier peak are shown in black while the rest is shown in grey. This is done for guidance to eye for separation of statistically significant result (black) and fitting procedure noise (grey).

4. Conclusions

An integrated photonic device specially designed to perform memory reading functionality is presented. The functionality is achieved through detection of the magnetization direction of an ultra-thin memory bit. The device is shown to operate despite performance reductions in the contributing building blocks and noise levels which correspond to ∼10 dB in terms of SNR. Post-processing of the intensity signal via Fourier transformation method stressed that the device is suitable as an analytical tool for research purposes. It is highlighted that the quantitative measurement of very small magneto-optic Kerr rotation (0.2°) is possible after a calibration which also considers optical loss.

Author Contributions: F.E.D. designed and directed the project, did FDTD simulations, contributed to the interpretation of the results and took the lead in writing the manuscript. C.v.d.B. developed the mathematical model, obtained and analyzed the results and prepared a draft manuscript. R.L. aided in interpreting the results, helped with methodology and provided input on the structure of the manuscript. J.J.G.M.v.d.T. and B.K. supervised the project, provided conceptualization of the work and reviewed/edited the work. All authors provided critical feedback and helped shape the research, analysis and manuscript. All authors have read and agreed to the published version of the manuscript.

Funding: This work is part of the Gravitation program 'Research Centre for Integrated Nanophotonics', which is financed by the Netherlands Organisation for Scientific Research (NWO).

Conflicts of Interest: The authors declare no conflict of interest.

References

1. Goodman, J.W.; Leonberger, F.J.; Kung, S.-Y.; Athale, R.A. Optical interconnections for VLSI systems. *Proc. IEEE* **1984**, *72*, 850–866. [CrossRef]
2. A Record Breaking Optical Chip. Available online: https://www.technologyreview.com/2008/06/25/219782/a-record-breaking-optical-chip/ (accessed on 11 November 2020).
3. Intel Leverages Chip Might Etch Photonics Future. Available online: https://www.nextplatform.com/2016/08/17/intel-leverages-chip-might-etch-photonics-future/ (accessed on 11 November 2020).
4. Kish, F.; Lal, V.; Evans, P.; Corzine, S.W.; Ziari, M.; Butrie, T.; Reffle, M.; Tsai, H.S.; Dentai, A.; Pleumeekers, J.; et al. System-on-chip photonic integrated circuits. *IEEE J. Sel. Top. Quantum Electron.* **2017**, *24*, 1–20. [CrossRef]
5. Sun, C.; Wade, M.T.; Lee, Y.; Orcutt, J.S.; Alloatti, L.; Georgas, M.S.; Waterman, A.S.; Shainline, J.M.; Avizienis, R.R.; Lin, S.; et al. Single-chip microprocessor that communicates directly using light. *Nature* **2015**, *528*, 534–538. [CrossRef] [PubMed]
6. Atabaki, A.H.; Moazeni, S.; Pavanello, F.; Gevorgyan, H.; Notaros, J.; Alloatti, L.; Wade, M.T.; Sun, C.; Kruger, S.A.; Meng, H.; et al. Integrating photonics with silicon nanoelectronics for the next generation of systems on a chip. *Nature* **2018**, *556*, 349–354. [CrossRef]
7. Smit, M.; Williams, K.; Van Der Tol, J. Past, present, and future of InP-based photonic integration. *APL Photonics* **2019**, *4*, 050901. [CrossRef]
8. Stanciu, C.; Hansteen, F.; Kimel, A.; Kirilyuk, A.; Tsukamoto, A.; Itoh, A.; Rasing, T. All-optical magnetic recording with circularly polarized light. *Phys. Rev. Lett.* **2007**, *99*, 047601. [CrossRef]
9. Lalieu, M.; Peeters, M.; Haenen, S.; Lavrijsen, R.; Koopmans, B. Deterministic all-optical switching of synthetic ferrimagnets using single femtosecond laser pulses. *Phys. Rev. B* **2017**, *96*, 220411. [CrossRef]
10. Van Hees, Y.; van der Tol, J.; Koopmans, B.; Lavrijsen, R. Periodically modulated ferromagnetic waveguide claddings with perpendicular magnetic anisotropy for enhanced mode conversion. *IEEE Photonics Benelux Proc.* **2017**, 196–199. Available online: http://www.photonics-benelux.org/images/stories/media/proceedings/2017/s17p196.pdf (accessed on 21 November 2020).
11. Kerr, J. On rotation of the plane of polarization by reflection from the pole of a magnet. *Lond. Edinb. Dublin Philos. Mag. J. Sci.* **1877**, *3*, 321–343. [CrossRef]
12. Freiser, M. A survey of magnetooptic effects. *IEEE Trans. Magn.* **1968**, *4*, 152–161. [CrossRef]
13. Parkin, S.S.; Hayashi, M.; Thomas, L. Magnetic domain-wall racetrack memory. *Science* **2008**, *320*, 190–194. [CrossRef]
14. Ryu, K.S.; Thomas, L.; Yang, S.H.; Parkin, S.S. Current induced tilting of domain walls in high velocity motion along perpendicularly magnetized micron-sized Co/Ni/Co racetracks. *Appl. Phys. Express* **2012**, *5*, 093006. [CrossRef]
15. Yang, S.H.; Ryu, K.S.; Parkin, S. Domain-wall velocities of up to 750 m s-1 driven by exchange-coupling torque in synthetic antiferromagnets. *Nat. Nanotechnol.* **2015**, *10*, 221–226. [CrossRef] [PubMed]
16. Caretta, L.; Mann, M.; Büttner, F.; Ueda, K.; Pfau, B.; Günther, C.M.; Hessing, P.; Churikova, A.; Klose, C.; Schneider, M.; et al. Fast current-driven domain walls and small skyrmions in a compensated ferrimagnet. *Nat. Nanotechnol.* **2018**, *13*, 1154–1160. [CrossRef] [PubMed]

17. Franken, J.; Swagten, H.; Koopmans, B. Shift registers based on magnetic domain wall ratchets with perpendicular anisotropy. *Nat. Nanotechnol.* **2012**, *7*, 499–503. [CrossRef] [PubMed]
18. Wang, M.; Zhang, Y.; Zhao, X.; Zhao, W. Tunnel junction with perpendicular magnetic anisotropy: Status and challenges. *Micromachines* **2015**, *6*, 1023–1045. [CrossRef]
19. Fiedler, S.; Stillrich, H.; Oepen, H.P. Magneto-optic properties of electron cyclotron resonance ion beam sputtered and magnetron sputtered Co/Pt multilayers. *J. Appl. Phys.* **2007**, *102*, 083906. [CrossRef]
20. Krinchik, G.S.; Artemjev, V.A. Magneto-optic Properties of Nickel, Iron, and Cobalt. *J. Appl. Phys.* **1968**, *39*, 1276–1278. [CrossRef]
21. Van der Tol, J.J.G.M.; Jiao, Y.; Van Engelen, J.P.; Pogoretskiy, V.; Kashi, A.A.; Williams, K. InP Membrane on Silicon (IMOS) Photonics. *IEEE J. Quantum Electron.* **2020**, *56*, 1–7. [CrossRef]
22. Lavrijsen, R.; Hartmann, D.M.F.; van den Brink, A.; Yin, Y.; Barcones, B.; Duine, R.A.; Verheijen, M.A.; Swagten, H.J.M.; Koopmans, B. Asymmetric magnetic bubble expansion under in-plane field in Pt/Co/Pt: Effect of interface engineering. *Phys. Rev. B* **2015**, *91*, 104414. [CrossRef]
23. Višňovský, Š.; Jakubisová Lišková, E.; Nývlt, M.; Krishnan, R. Origin of magneto-optic enhancement in CoPt alloys and Co/Pt multilayers. *Appl. Phys. Lett.* **2012**, *100*, 232409. [CrossRef]
24. Lumerical Solutions. Available online: https://www.lumerical.com/products/ (accessed on 29 October 2020).
25. Atkinson, R.; Pahirathan, S.; Salter, I.W.; Grundy, P.J.; Tatnall, C.J.; Lodder, J.C.; Meng, Q. Fundamental optical and magneto-optical constants of CoPt and CoNiPt multilayered films. *J. Magn. Magn. Mater.* **1996**, *162*, 131–138. [CrossRef]
26. Pello, J.; van der Tol, J.; Keyvaninia, S.; van Veldhoven, R.; Ambrosius, H.; Roelkens, G.; Smit, M. High-efficiency ultrasmall polarization converter in InP membrane. *Opt. Lett.* **2012**, *37*, 3711–3713. [CrossRef] [PubMed]

Publisher's Note: MDPI stays neutral with regard to jurisdictional claims in published maps and institutional affiliations.

© 2020 by the authors. Licensee MDPI, Basel, Switzerland. This article is an open access article distributed under the terms and conditions of the Creative Commons Attribution (CC BY) license (http://creativecommons.org/licenses/by/4.0/).

Article

Theoretical Investigation of Responsivity/NEP Trade-off in NIR Graphene/Semiconductor Schottky Photodetectors Operating at Room Temperature

Teresa Crisci [1,2], Luigi Moretti [1] and Maurizio Casalino [2,*]

[1] Department of Mathematics and Physics, University of Campania "Luigi Vanvitelli", Viale Abramo Lincoln, 5, 81100 Caserta, Italy; teresa.crisci@na.isasi.cnr.it (T.C.); luigi.moretti@unicampania.it (L.M.)
[2] Institute of Applied Science and Intelligent Systems "Eduardo Caianiello" (CNR), Via Pietro Castellino, 111, 80131 Naples, Italy
* Correspondence: maurizio.casalino@na.isasi.cnr.it; Tel.: +39-081-6132345

Abstract: In this work we theoretically investigate the responsivity/noise equivalent power (NEP) trade-off in graphene/semiconductor Schottky photodetectors (PDs) operating in the near-infrared regime and working at room temperature. Our analysis shows that the responsivity/NEP ratio is strongly dependent on the Schottky barrier height (SBH) of the junction, and we derive a closed analytical formula for maximizing it. In addition, we theoretically discuss how the SBH is related to the reverse voltage applied to the junction in order to show how these devices could be optimized in practice for different semiconductors. We found that graphene/n-silicon (Si) Schottky PDs could be optimized at 1550 nm, showing a responsivity and NEP of 133 mA/W and 500 fW/$\sqrt{Hz}$, respectively, with a low reverse bias of only 0.66 V. Moreover, we show that graphene/n-germanium (Ge) Schottky PDs optimized in terms of responsivity/NEP ratio could be employed at 2000 nm with a responsivity and NEP of 233 mA/W and 31 pW/$\sqrt{Hz}$, respectively. We believe that our insights are of great importance in the field of silicon photonics for the realization of Si-based PDs to be employed in power monitoring, lab-on-chip and environment monitoring applications.

Keywords: graphene; silicon; photodetectors; internal photoemission effect; near-infrared

1. Introduction

Silicon (Si) Schottky photodetectors (PDs) have attracted the interest of the scientific community due to the possibility of making Si suitable for detecting infrared (IR) radiation, which is the range of wavelengths included in the spectrum where Si has a negligible optical absorption due to its bandgap of 1.12 eV (1.1 µm). Schottky Si PDs are metal/Si junctions whose detection mechanism is based on the internal photoemission effect (IPE), that is, the photo-excitation of charge carriers in the metal and their emission into Si over the Schottky barrier of the junction [1–3]. In other words, in Si Schottky PDs the metal and not the Si is the active material absorbing the incoming optical radiation. In this context, both palladium silicide (Pd_2Si) and platinum silicide (PtSi) Schottky PDs have been extensively investigated for the realization of infrared CCD image sensors. Pd_2Si/Si Schottky PDs were developed for satellite applications showing the ability to detect a spectrum ranging from 1 to 2.5 µm when cooled to a temperature of 120 K [4,5]. On the other hand, PtSi/Si Schottky PDs were developed for operation at longer wavelengths ranging from 3 to 5 µm [6,7], although they require a lower temperature of 80 K. A focal plane array (FPA) constituted by an array of 512 × 512 PtSi/Si pixels was realized, demonstrating the first spectacular convergence between Si photonics and electronics [8]. Unfortunately, these devices can only work at cryogenic temperature. Indeed, the low Schottky barrier height (SBH) required to achieve an acceptable efficiency (0.21 eV for PtSi [7] and 0.34 eV for Pd_2Si [4]) is comes at the cost of PD noise (dark current), which must be reduced by lowering the working temperature. PD noise affects the noise equivalent power (NEP), that is, the minimum detectable optical power, which has a huge impact on both the device sensitivity and the bit error rate (BER) of a communication link. Higher Schottky barriers make it possible to achieve low noise, but they unfortunately also lead to low efficiencies. This efficiency–noise trade-off is a peculiar characteristic of the Schottky PDs based on the IPE.

In 2006, for the first time, it was theoretically proposed to use Schottky PDs for the detection of near-IR (NIR) wavelengths at room temperature [9], taking advantage of the interference phenomena occurring inside a high-finesse Fabry–Pérot microcavity. The main idea was to work with metal/semiconductor junctions characterized by higher SBHs in order to reduce the dark current and then to recover the device efficiency by increasing the metal absorption through the multiple reflections of the optical radiation inside the microcavity. Later, many other strategies were pursued to enhance the efficiency of these devices; indeed, surface plasmon polaritons (SPPs) [10,11], Si nanoparticles (NPs) [12], metallic antennas [13], and gratings [14] were proposed and investigated. In any case, the measured responsivity was lower than 30 mA/W [12] and 5 mA/W [15] for waveguide and free-space Schottky PDs, respectively. More important, the efficiency–noise trade-off of these Schottky PDs has never been optimized in terms of SBH for achieving high efficiency and low noise at the same time. The low responsivity (i.e., the ratio between the photogenerated current and the incoming optical power) of the Schottky PDs based on metals is mainly due to the small emission probability of the photo-excited carriers from the metal to the Si, related to the momentum mismatch.

Recently, graphene/Si Schottky PDs have shown higher efficiencies with respect to the metallic counterpart and, even if the physical mechanism behind this enhancement is still under debate, it seems related to the increased emission probability due to the two-dimensionality of the material [16–18]. Although graphene is characterized by a low optical absorption (2.3%) many approaches based on resonant-cavity-enhanced (RCE) configurations [19,20], plasmonic structures [21], waveguiding structures [22], and quantum dots [23] have been proposed to overcome this drawback. At present, graphene/Si PDs [18,22,24] show superior performance to the corresponding metallic PDs, representing the most promising solution to realize low-cost Si PDs operating in the NIR regime. In addition, graphene offers a novel attractive possibility: the graphene Fermi level (i.e., the SBH with Si), can be simply modified by applying a bias to the junction, making it feasible to optimize the efficiency–noise trade-off.

In this work we theoretically investigated the responsivity/NEP trade-off in graphene/semiconductor Schottky PDs operating at NIR wavelengths and at room temperature. First, we used the results of the recent literature to derive a responsivity/NEP analytical equation that can be maximized with an appropriate choice of SBH. Then, we reviewed the SBH dependence on the bias applied to the graphene/semiconductor junctions to show how the responsivity/NEP ratio could be maximized in practice. Finally, we numerically calculated both the responsivity and the NEP of graphene/semiconductor PDs discussing their possible applications and highlighting the validity limits of the proposed optimization process. Even if this work was carried out with the aim of gaining greater insight into graphene/Si PDs, it is worth mentioning that we trace here a general methodology which can also be applied to different semiconductors, such as: germanium (Ge), gallium arsenide (GaAs), and aluminum gallium arsenide (AlGaAs).

2. Theoretical Background

IPE theory was first developed by Fowler in 1931, and it was focused on the injection of electrons from a metal into vacuum [25]. Several authors have extended Fowler's theory to the emission of carriers into semiconductors, conceiving the modified Fowler theory [26–28] and providing the following expression for the internal quantum efficiency (IQE) η_{int} of IPE-based PDs, defined as the number of charge carriers N_e produced per absorbed photons N_{ass} [26]:

$$\eta_{int} = \frac{N_e}{N_{ass}} = \frac{1}{8E_F} \cdot \frac{(h\nu - q\Phi_B)^2}{h\nu} \qquad (1)$$

where E_F represents the Fermi level, $h\nu = hc/\lambda$ is the energy of the incident photon (λ is the wavelength and c the speed of light in a vacuum), q is the electron charge, and Φ_B is the potential barrier at the interface between the metal and the semiconductor. This expression is derived by taking into account the ratio of charge carriers having kinetic energy *normal* to the surface of the junction, necessary to overcome the potential barrier. This mechanism usually leads to poor efficiency (about 1%) [29,30]; however, it has been demonstrated that two-dimensional materials replacing metals in the Schottky junctions provide an IQE enhancement [18]. In particular, in single-layer graphene (SLG)/semiconductor junctions a still higher ratio of photon conversion in charge carriers is observed. Regarding this, Amirmazlaghani et al. [18] explain how this can be ascribed to the molecular structure of the graphene. Indeed, the π orbitals are normal to the interface with the semiconductor, and the charge carriers' momentum can be directed only towards the semiconductor or in the opposite direction, leading to an enhancement of the emission probability up to $\frac{1}{2}$. When SLG is used as active medium in an IPE-based PD, Equation (1) can no longer be applied due to the linearity of the dispersion relation near the Dirac point [31], different density of states, and probability of emission. However, the IQE of Schottky PDs based on SLG has been derived as [18]:

$$\eta_{int}^{SLG} = \frac{1}{2} \cdot \frac{(h\nu)^2 - (q\Phi_B)^2}{(h\nu)^2}. \qquad (2)$$

The responsivity R is related to η_{int}^{SLG} by the following relation:

$$R = \frac{I_{ph}}{P_{inc}} = S \cdot \frac{1}{h\nu} \cdot \eta_{int}^{SLG} = \frac{S}{2} \cdot \frac{(h\nu)^2 - (q\Phi_B)^2}{(h\nu)^3} \qquad (3)$$

where I_{ph} is the photogenerated current, P_{inc} is the incident optical power, and S is the graphene optical absorbance. It is worth mentioning that in Equation (3) the charge carrier q is been considered in order to express the responsivity in A/W. Graphene has an optical absorption related to the universal fine-structure constant $\alpha = e^2/(\pi \epsilon_0 \hbar c)$ [32] and independent of the frequency, $A_G = \pi \alpha \approx 2.3\%$. Here we focus our attention on devices that provide the complete absorption of the incident radiation such as long waveguides and resonant structures, thus we consider $S = 1$.

As the Schottky barrier Φ_B decreases, more electrons can pass into the semiconductor, giving rise to higher responsivities, as shown in Equation (2). Unfortunately, the dark current I_d of the junction also increases as Φ_B diminishes due to thermal effects [33]:

$$I_d = A_j A^* T^2 \cdot e^{-\frac{q\Phi_B}{kT}} \qquad (4)$$

where A_j is the area of the Schottky junction, A^* is the Richardson constant, T is the absolute temperature and k is the Boltzmann constant. Furthermore, there is a component of noise intrinsic to the photodetection mechanism: due to the quantized nature of the light, the current is constituted by a succession of random impulses, which cause fluctuations of the measured current (shot noise). The quadratic mean value of the fluctuations linked to both photocurrent I_{ph} and dark current I_d is the following:

$$i_s^2(\Phi_B) = 2qB(I_d(\Phi_B) + I_{ph}(\Phi_B)) \qquad (5)$$

where B is the device bandwidth. In addition to the shot noise, there is a thermal noise (Johnson noise) with quadratic mean value:

$$i_R^2 = \frac{4kTB}{R_L}, \qquad (6)$$

where R_L is the load resistance of the PD. Since the two contributions of the noise current are statistically independent, the total noise i_n is given by their squared sum:

$$i_n = \sqrt{2qB(I_d(\Phi_B) + I_{ph}(\Phi_B)) + \frac{4kTB}{R_L}}. \quad (7)$$

At low signal levels $I_{ph} \ll I_d$, the condition to make the thermal noise negligible compared to the shot noise in Equation (7) is:

$$I_d \gg 2V_{th}/R_L, \quad (8)$$

where the thermal voltage $V_{th} = kT/q$. At room temperature, Equation (8) mainly depends on both SBH and R_L. Of course, if the thermal noise dominates the shot noise, i_n does not depend on the SBH and the optimization procedure reported here can no longer be adopted. Compared to the absolute value of i_n, its magnitude compared to the generated signal I_{ph}, defined as the signal-to-noise ratio $SNR = I_{ph}/i_n$, is even more important.

In order to find the value of photogenerated current I_{ph} that brings $SNR = 1$, we can take advantage of the definition of the SNR and considering Equations (7) and (8), we obtain:

$$SNR = \frac{I_{ph}}{\sqrt{2qB(I_d(\Phi_B) + I_{ph}(\Phi_B))}} = 1. \quad (9)$$

The square of the previous equation gives a quadratic form in the unknown I_{ph}; by solving it we find:

$$I_{ph} = qB\left(1 \pm \sqrt{1 + \frac{2I_d}{qB}}\right). \quad (10)$$

This expression makes it possible to obtain the minimum incident optical power P_{inc} necessary to get $SNR = 1$ for a PD characterized by a responsivity R. Since the NEP is defined as the incident optical power P_{inc} necessary to get $SNR = 1$ divided by the square root of the bandwidth ($NEP = P_{inc}/\sqrt{B}$), we numerically obtain NEP by considering $B = 1$ Hz in Equation (10) and dividing it by the responsivity R:

$$NEP = \frac{q\left(1 \pm \sqrt{1 + \frac{2I_d}{q}}\right)}{R} \quad (11)$$

which reduces to the very well-known formula:

$$NEP \approx \frac{\sqrt{2qI_d}}{R}, \quad (12)$$

where $2I_d/q$ is much larger than 1 in typical PDs. It is worth noting that in Equation (12) the sign of R follows the sign of I_{ph}, as is clear when looking at Equation (3).

Optimized PDs are characterized by high responsivity and low NEP. However, by looking at Equations (3) and (12) it is clear that by increasing the SBH, the NEP improves at the expense of the responsivity. On the other hand, an SBH decrease is beneficial in terms of responsivity but it degrades the NEP. Hence, we sought investigate the Schottky barrier Φ_B that maximizes the R to NEP ratio. Toward this aim we introduce the function $G(\Phi_B) = \sqrt{\frac{R}{NEP}}$ using Equations (2)–(4), and (12):

$$G(\Phi_B) = \sqrt{\frac{R}{NEP}} = \frac{R}{\sqrt[4]{2qI_d}} = C \cdot \frac{(h\nu)^2 - (q\Phi_B)^2}{\sqrt{T}(h\nu)^3} \cdot e^{\frac{q\Phi_B}{4kT}} = C \cdot g(\Phi_B) \quad (13)$$

where $C = 1/(2\sqrt[4]{2qA_jA^*})$ depends on the geometry through the junction area A_j and on the semiconductor through the Richardson constant A^*. Figure 1a displays the behavior of $g(\Phi_B)$ at 300 K for three different wavelengths, 1.3 µm, 1.55 µm, and 2 µm, showing the presence of a peak. By calculating the first and second derivatives of $G(\Phi_B)$ we can find the value Φ_B^* of SBH corresponding to this peak:

$$\Phi_B^* = -4kT\left[1 - \sqrt{1 + \frac{(h\nu)^2}{16(kT)^2}}\right]. \tag{14}$$

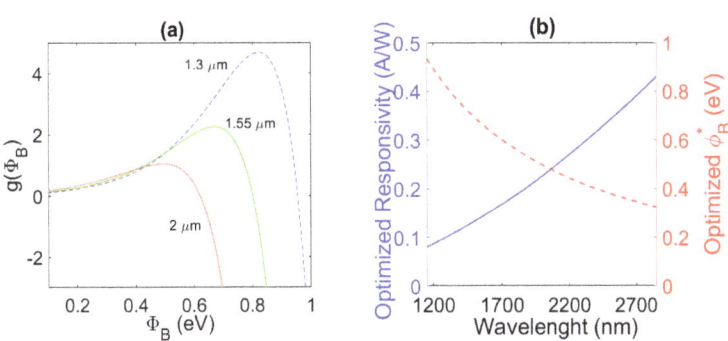

Figure 1. (a) Behavior of $g(\Phi_B)$ at 300 K for three wavelengths: 1.3 µm, 1.55 µm, and 2 µm; (b) optimized responsivity R (blue solid line) and optimized Schottky barrier height (SBH) Φ_B^* (red dashed line) as a function of the wavelengths.

We define Φ_B^* as the optimized SBH because it is the value at which the R-to-NEP ratio is maximized. The dashed red line in Figure 1b shows the optimized Schottky barrier Φ_B^* as a function of the wavelength. This behavior can be explained by considering that when the wavelength is reduced the photon energy $h\nu$ increases by diminishing the responsivity R, as shown in Equation (3), requiring a reduction of the NEP to maintain the maximized R/NEP ratio. In turn, the NEP reduction can be achieved by an increase of the optimized Φ_B^*, which decreases the amount of charge carriers able to overcome the Schottky barrier due to thermal effects. Even if the Φ_B^* increment also produces a decrease in responsivity, it is important to recall that while the NEP is characterized by an exponential decay as a function of Φ_B^* ($NEP \sim e^{-\frac{\Phi_B^*}{2V_t}}$), the responsivity is characterized by a simple quadratic behaviour ($R \sim \Phi_B^{*2}$).

The substitution of Equation (14) in Equation (3) provides the responsivity when the ratio R/NEP is maximized (here we refer to it as the optimized responsivity), as shown by the blue solid line in Figure 1b. Note that this optimized responsivity depends only on the SBH of the junction. Figure 1b shows how the optimized responsivity is increased by increasing the wavelength owing to a drop in the optimized SBH Φ_B^*, providing values at room temperature of 0.10 A/W, 0.14 A/W, and 0.23 A/W at 1.3 µm, 1.55 µm, and 2 µm, respectively, as reported in Table 1. If higher responsivities are required, they can be achieved by lowering the SBH but at the expense of the SNR.

Table 1. Values of the Schottky barrier Φ_B^* optimizing the responsivity (R)/noise equivalent power (NEP) ratio at the three wavelengths of interest: 1.3 µm, 1.55 µm, and 2 µm at $T = 300$ K. The corresponding efficiency η_{int}^{SLG} and responsivity R, calculated respectively through Equations (2), (3) and (14), are also shown. SLG is the acronym of single layer graphene.

λ (µm)	Φ_B^* (eV)	η_{int}^{SLG}	R (A/W)
1.3	0.86	0.10	0.10
1.55	0.7	0.11	0.14
2	0.52	0.14	0.23

3. Theoretical Results and Discussion

In this section we theoretically derive the SBH dependence on the bias applied to the junction in order to show how the graphene Schottky PDs based on different semiconductors could be optimized.

It is well-known that the SBH Φ_B of Schottky PDs can be determined by the two following equations (i.e., the Schottky–Mott relations) [33]:

$$q\Phi_B^{(n)}(V_R) = q\Phi_{gr}^0 - \Delta E_F(V_R) - q\chi_{sm} \quad \text{(n-type)} \tag{15}$$

$$q\Phi_B^{(p)}(V_R) = E_g - (q\Phi_{gr}^0 - \Delta E_F(V_R) - q\chi_{sm}) \quad \text{(p-type)} \tag{16}$$

where χ_{sm} and E_g are, respectively, the electron affinity and the bandgap of the semiconductor and $q\Phi_{gr}^0$ is the difference between the vacuum level E_0 and the Dirac point E_F^0, while the graphene Fermi level is E_F (Figure 2). Therefore, $\Delta E_F = E_F - E_F^0$ can be expressed as [34]:

$$\Delta E_F = -sgn(n)\hbar v_F \sqrt{\pi|n|} \tag{17}$$

where $v_F = 1.1 \times 10^8$ cm/s is the Fermi velocity, $\hbar$ is the reduced Planck constant, and n is the carrier density in graphene. The carrier density n not only depends on the graphene extrinsic doping n_0 (defined positive and negative for p-type and n-type graphene doping, respectively) but also on the thermal contact with the semiconductor. Indeed, when a p-doped graphene ($n_0 > 0$) is transferred onto the semiconductor, the space charge Q_{sm} in the depletion region induces an opposite charge $Q_{gr} = -Q_{sm}$ in the graphene layer. This creates additional charge carriers, modifying the carrier density, which becomes $n = n_0 + \frac{Q_{gr}}{q}$. The expression of the space charge Q_{sm} when the region is completely depleted is $Q_{sm} = \pm\sqrt{2\epsilon_{sm}NqV_{bi}}$, where ϵ_{sm} and N are the dielectric permittivity and the doping density of the semiconductor, respectively, while V_{bi} is the built-in potential. Moreover, by applying a reverse voltage, the charge per unit area in the graphene becomes $Q_{gr} = \mp\sqrt{2\epsilon_{sm}Nq(V_{bi} + V_R)}$, providing a carrier density:

$$n = n_0 \mp \sqrt{\frac{2\epsilon_{sm}}{q}N(V_{bi} + V_R)} \tag{18}$$

where the signs minus and plus are for n- and p-type semiconductors, respectively. Equation (18) replaced into Equation (17) and then in Equation (15) or (16) gives the desired dependence between the SBH and the reverse bias V_R.

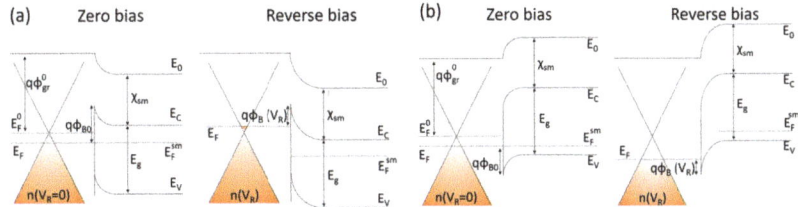

Figure 2. Band diagrams of (**a**) graphene/n-semiconductor and (**b**) graphene/p-semiconductor junctions at the thermal equilibrium and when a reverse bias V_R is applied. At the thermal equilibrium, graphene has an initial carrier density $n(V_R = 0)$. After a reverse bias this charge density becomes $n(V_R)$. E_0 represents the vacuum energy level while E_F^0 is the Dirac point. Φ_{gr}^0, χ_{sm}, E_g, E_C, and E_V are respectively the intrinsic graphene work function, electron affinity, conduction band, bandgap, and valence band. E_F^{sm} is the Fermi energy level in the semiconductor and $q\Phi_{B0}$ the Schottky barrier at zero bias. The values of the Schottky barrier $q\Phi_B$ depend on the graphene Fermi energy level E_F that shifts when a voltage is applied.

In Table 2 we report the bandgap energy and the electron affinity for various semiconductors and the SBH at zero bias Φ_{B0}, calculated through Equation (15) or (16) when $V_R = 0$. The values of Φ_{B0} were evaluated by considering a graphene work function $\Phi_{gr}^0 = 4.6$ eV [35,36], a built-in potential $V_{bi} = 0.6$ V, an initial SLG extrinsic p-doping $n_0 = 10^{12}$ cm^{-2}, and a low doping of the semiconductors $N = 10^{16}$ cm^{-3}.

Table 2. Bandgap E_g and electron affinity χ_{sm} of various semiconductors together with values of SBH when the Schottky junction is formed, calculated thanks to Equations (15)–(17) by taking into account an initial extrinsic p-doping $n_0 = 10^{12}$ cm^{-2} of the single-layer graphene (SLG) and the thermal equilibrium contact with the substrate. For the calculations we considered low-doped semiconductors (i.e., $N = 10^{16}$ cm^{-3}).

Semiconductor	E_g (eV)	χ_{sm} (eV)	$\Phi_{B0}^{(n)}$ (eV)	$\Phi_{B0}^{(p)}$ (eV)
Si	1.12	4.00	0.73	0.39
GaAs	1.43	4.07	0.66	0.77
$Al_{0.3}Ga_{0.7}As$	1.77	3.77	0.96	0.84
Ge	0.66	4.13	0.60	—

Figure 3a shows the intersections between these values of SBH Φ_{B0} for different semiconductors and the curve of the optimized $\Phi_B^*(\lambda)$ at room temperature (given by Equation (14)), suggesting the working wavelength to achieve the highest R/NEP ratio for each material. In the range of wavelengths where $\Phi_{B0} > \Phi_B^*(\lambda)$, the SBH can be lowered down to its optimal value as in Equation (14) by simply applying a specific reverse bias V_R to the junction.

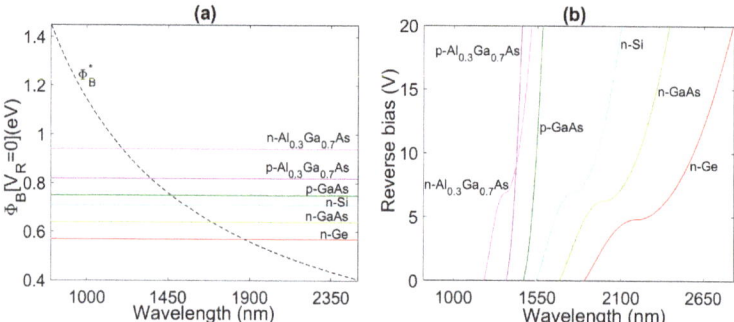

Figure 3. (a) Intersection between the curve $\Phi_B^*(\lambda)$ at 300 K and the values of SBH Φ_{B0} at the interface between graphene and several semiconductors in conditions of thermal contact (no voltage applied to the junction); (b) reverse voltage V_R to apply to the graphene/semiconductor junction as function of the wavelength for maximizing the signal-to-noise ratio (SNR) ($\Phi_B = \Phi_B^*$) for various semiconductors. The values of $\Phi_{B0} = \Phi_B(V_R = 0)$ were calculated through Equations (15)–(18) by considering an initial graphene p-doping of $n_0 = 10^{12}$ cm^{-2} and a doping of $N = 10^{16}$ cm^{-3} for all the semiconductors reported in Table 2.

By inverting Equation (18) and using Equation (17) and Equation (15) or (16) it is possible to calculate, for each wavelength and for each semiconductor, the values of the reverse voltage V_R such that $\Phi_B = \Phi_B^*$. We report this plot in Figure 3b by considering a maximum reverse bias of 20 V. It is interesting to observe that within this limit, graphene Schottky PDs based on p-Al$_{0.3}$Ga$_{0.7}$As and p-GaAs can be optimized only in a narrow window of the NIR spectrum, whereas n-Si can be optimized in a broader range, including at 1.55 μm where only a small reverse voltage $V_R = 0.66$ V for maximizing the R/NEP ratio is required. Indeed, at a reverse voltage of 0.66 V the $\Phi_{B0} = 0.73$ eV of the graphene/n-Si junction can be reduced to its optimum value of $\Phi_B^*(1.55\,\mu m) = 0.71$ eV. In contrast, p-GaAs requires a higher reverse voltage of 12 V to maximize R/NEP. Finally, n-Ge stands out among the analyzed semiconductors in view of the possibility to be employed over a region of the NIR spectrum above 2 μm. The range of wavelengths where R/NEP can be optimized for various semiconductors, by applying a reverse bias up to 20 V, is summarized in Table 3.

Table 3. Range of wavelengths in which the R/NEP ratio of the Schottky photodetectors (PDs) can be maximized by applying a reverse bias up to 20 V.

Semiconductor	λ_{min} (nm)	λ_{max} (nm)
$n - Si$	1541	2099
$p - GaAs$	1459	1582
$n - GaAs$	1692	2417
$p - Al_{0.3}Ga_{0.7}As$	1346	1447
$n - Al_{0.3}Ga_{0.7}As$	1197	1508
$n - Ge$	1852	2843

In Figure 4a,b we report the values of the quantities of interest in this work—the R/NEP ratio and the optimized NEP—for all the examined semiconductors by considering a graphene circular area with radius of 500 μm and a PD closed on a load resistance of 10 MΩ. We compute these optimized quantities through Equation (14) substituted into Equations (3) and (12). Recall that the results shown in Figure 4 are valid when the condition in Equation (8) is fulfilled. In order to verify it, we consider the dark current I_d one order of magnitude higher than $\frac{2V_{th}}{R_L}$ ($I_d = 10\frac{2V_{th}}{R_L}$), and we calculate both optimized R/NEP and NEP by Equation (12). The solid black lines drawn in Figure 4a,b represent the validity thresholds of our discussion: graphene Schottky PDs can be optimized in terms of R/NEP ratio at a given wavelength by means the use of semiconductors placed below and above the solid black lines drawn in Figure 4a,b, respectively. These thresholds depend on the load resistance R_L, the SBH Φ_B, and the graphene active area A_j, as is clearly shown by Equation (8). We discover that in the case analyzed here, only graphene/n-Si, graphene/n-Ge, and graphene/n-GaAs Schottky PDs can be suitable for this optimization procedure. Although Si is typically used for visible detection, analysis shows that graphene/n-Si Schottky PDs with a maximized R/NEP ratio could be adopted for detecting sub-bandgap NIR wavelengths with responsivity and NEP of 133 mA/W and 500 fW/$\sqrt{Hz}$ at 1.55 µm, respectively. These devices provide low NEP, enabling their employment for power monitoring and lab-on-chip applications. Note that the predicted responsivity of graphene/n-Si PDs is higher than that reported for NIR Si PDs based on bulk-defect-mediated absorption. Indeed, taking advantage of mid-gap defects introduced into Si ring and disk resonators, Ackert et al. reported a responsivity of only 23 mA/W at −5 V [37] and 45 mA/W at −3 V [38] at 1560 nm, respectively. On the other hand, if the inter-band absorption of Ge is typically used for detecting the wavelength of 1.55 µm for telecommunications applications, graphene/n-Ge Schottky PDs could allow the detection of wavelengths longer than 1.55 µm, where the the Ge inter-band absorption suddenly decreases. Indeed, graphene/n-Ge Schottky PDs with optimized R/NEP ratio show a responsivity and NEP of 227 mA/W and 31 pW/$\sqrt{Hz}$ at 2 µm, respectively, enabling their employment in environment monitoring applications. The predicted responsivity of graphene/n-Ge PDs is higher than that reported for NIR Ge PDs based on the introduction of tin (Sn) atoms in the Ge lattice. Indeed, with a substitutional Sn concentration of 6.5%, Ge-based PDs are able to absorb optical radiation at 2 µm but provide a limited responsivity of only 20 mA/W [39]. Note that NEP depends on many parameters, such as graphene's optical absorbance, the graphene area in contact with the semiconductor, and the temperature. Among these, particular attention should be paid to the temperature, which appears in the exponential argument of the dark current (Equation (4)), which in turn affects the NEP (Equation (12)). As an example, in a graphene/n-Si Schottky PD we evaluated this by increasing the temperature of 1 °C with respect to the room temperature, and an increase in optimized NEP below 5% could be achieved at any wavelength in the range of interest for this junction. As reported in Figure 4a,b, semiconductors such as n-Si, n-Ge, and n-GaAs can be exploited at room temperature for the realization of optimized graphene-based Schottky PDs in the spectral range from 1955 to 2080 nm with a responsivity from 219 to 245 mA/W (Figure 1b); however, while n-GaAs would be characterized by a lower NEP, n-Si and n-Ge would have the advantage of a better compatibility with CMOS technology.

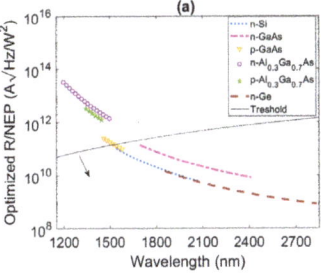
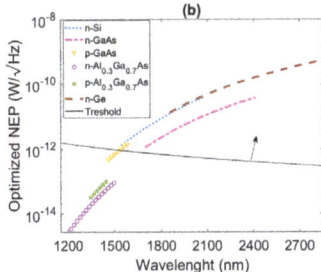

Figure 4. (a) The optimized R/NEP and (b) the optimized NEP of the Schottky graphene-based PDs for various semiconductors as function of the wavelength range individuated in Table 3. All figures were obtained at room temperature and by considering a graphene circular area in touch with the semiconductor with radius of 500 µm and a load resistance of 10 MΩ. The arrows indicate the validity regions of the proposed optimization procedure.

4. Conclusions

In this work we theoretically investigated the responsivity/NEP trade-off of NIR graphene/semiconductor Schottky PDs at room temperature. An analytical expression of the SBH able to maximize the R/NEP ratio was derived. Furthermore, we discussed how the optimized SBH can be tuned by applying a reverse voltage to the junction in order to establish the best operation conditions to achieve higher responsivity as well as lower noise for various semiconductors. Toward this aim we have accounted for the physics behind the emission of photo-excited charge carriers from graphene to Si, the theory of the graphene/semiconductor Schottky junctions, and the properties of graphene related to its two-dimensionality.

Remarkably, we found that CMOS-compatible materials such as Si and Ge could be exploited for the realization of optimized graphene Schottky PDs able to detect wavelengths beyond the limit imposed by their inter-band optical absorption. Indeed, graphene/n-Si Schottky PDs with maximized R/NEP ratio showed responsivity and NEP of 133 mA/W and 500 fW/$\sqrt{Hz}$, respectively, at 1.55 μm by applying a reverse voltage of only 0.66 V. On the other hand, graphene/n-Ge Schottky PDs with maximized R/NEP ratio showed the potential to work at wavelengths longer than 1.55 μm, being for instance characterized by a responsivity and NEP of 227 mA/W and 31 pW/$\sqrt{Hz}$ at 2 μm.

We believe that the insights reported in this work could be of paramount importance in silicon photonics for the realization of optimized PDs to be employed in power monitoring, lab-on-chip, and environment monitoring applications.

Author Contributions: Conceptualization, T.C. and M.C.; methodology, T.C., M.C., and L.M.; formal analysis, T.C. and M.C.; resources, L.M.; data curation, T.C., M.C. and L.M.; writing original draft preparation, T.C.; writing—review and editing, M.C. and L.M. All authors have read and agreed to the published version of the manuscript.

Funding: This research received no external funding.

Institutional Review Board Statement: Not applicable.

Informed Consent Statement: Not applicable.

Conflicts of Interest: The authors declare no conflict of interest.

References

1. Casalino, M. Internal photoemission theory: Comments and theoretical limitations on the performance of near-infrared silicon Schottky photodetectors. *IEEE J. Quantum Electron.* **2016**, *52*, 1–10. [CrossRef]
2. Scales, C.; Berini, P. Thin-film Schottky barrier photodetector models. *IEEE J. Quantum Electron.* **2010**, *46*, 633–643. [CrossRef]
3. Crisci, T.; Moretti, L.; Gioffrè, M.; Iodice, M.; Coppola, G.; Casalino, M. Integrated Er/Si Schottky Photodetectors on the end facet of optical waveguides. *J. Eur. Opt. Soc. Rapid Publ.* **2020**, *16*, 1–8. [CrossRef]
4. Elabd, H.; Villani, T.; Kosonocky, W. Palladium-silicide Schottky-barrier IR-CCD for SWIR applications at intermediate temperatures. *IEEE Electron Device Lett.* **1982**, *3*, 89–90. [CrossRef]
5. Elabd, H.; Villani, T.; Tower, J. High density Schottky barrier IRCCD sensors for SWIR applications at intermediate temperature. In Proceedings of the Society of Photo-Optical Instrumentation Engineers (SPIE) Conference, Arlington, TX, USA, November 1982; pp. 161–171.
6. Kosonocky, W.; Elabd, H.; Erhardt, H.; Shallcross, F.; Villani, T.; Meray, G.; Cantella, M.; Klein, J.; Roberts, N. 64× 128-Element high-performance PtSi IR-CCD imager sensor. In Proceedings of the International Electron Devices Meeting, Washington, DC, USA, 7–9 December 1981; p. 702. Available online: https://ntrs.nasa.gov/citations/19830020268 (accessed on 9 April 2021).
7. Kosonocky, W.; Elabd, H.; Erhardt, H.; Shallcross, F.; Meray, G.; Villani, T.; Groppe, J.; Miller, R.; Frantz, V.; Cantella, M. Design And Performance of 64 × 128 Element PtSi Schottky-Barrier Infrared Charge-Coupled Device (IRCCD) Focal Plane Array. In Proceeding of the Society of Photo-Optical Instrumentation Engineers (SPIE), Infrared Sensor Technology, Arlington, USA, 28 December 1982; Volume 344, pp. 66–77.
8. Wang, W.L.; Winzenread, R.; Nguyen, B.; Murrin, J.J.; Trubiano, R.L. High fill factor 512 × 512 PtSi focal plane array. In *New Methods in Microscopy and Low Light Imaging*; International Society for Optics and Photonics: San Diego, CA, USA, 1989; Volume 1161, pp. 79–95.
9. Casalino, M.; Sirleto, L.; Moretti, L.; Della Corte, F.; Rendina, I. Design of a silicon resonant cavity enhanced photodetector based on the internal photoemission effect at 1.55 μm. *J. Opt. A Pure Appl. Opt.* **2006**, *8*, 909. [CrossRef]

10. Berini, P.; Olivieri, A.; Chen, C. Thin Au surface plasmon waveguide Schottky detectors on p-Si. *Nanotechnology* **2012**, *23*, 444011. [CrossRef]
11. Akbari, A.; Tait, R.N.; Berini, P. Surface plasmon waveguide Schottky detector. *Opt. Express* **2010**, *18*, 8505–8514. [CrossRef]
12. Zhu, S.; Chu, H.; Lo, G.; Bai, P.; Kwong, D. Waveguide-integrated near-infrared detector with self-assembled metal silicide nanoparticles embedded in a silicon pn junction. *Appl. Phys. Lett.* **2012**, *100*, 061109. [CrossRef]
13. Knight, M.W.; Sobhani, H.; Nordlander, P.; Halas, N.J. Photodetection with active optical antennas. *Science* **2011**, *332*, 702–704. [CrossRef]
14. Sobhani, A.; Knight, M.W.; Wang, Y.; Zheng, B.; King, N.S.; Brown, L.V.; Fang, Z.; Nordlander, P.; Halas, N.J. Narrowband photodetection in the near-infrared with a plasmon-induced hot electron device. *Nat. Commun.* **2013**, *4*, 1–6. [CrossRef]
15. Desiatov, B.; Goykhman, I.; Mazurski, N.; Shappir, J.; Khurgin, J.B.; Levy, U. Plasmonic enhanced silicon pyramids for internal photoemission Schottky detectors in the near-infrared regime. *Optica* **2015**, *2*, 335–338. [CrossRef]
16. Levy, U.; Grajower, M.; Goncalves, P.; Mortensen, N.A.; Khurgin, J.B. Plasmonic silicon Schottky photodetectors: The physics behind graphene enhanced internal photoemission. *APL Photonics* **2017**, *2*, 026103. [CrossRef]
17. Casalino, M.; Russo, R.; Russo, C.; Ciajolo, A.; Di Gennaro, E.; Iodice, M.; Coppola, G. Free-space schottky graphene/silicon photodetectors operating at 2 µm. *ACS Photonics* **2018**, *5*, 4577–4585. [CrossRef]
18. Amirmazlaghani, M.; Raissi, F.; Habibpour, O.; Vukusic, J.; Stake, J. Graphene-Si Schottky IR Detector. *IEEE J. Quantum Electron.* **2013**, *49*, 589–594. [CrossRef]
19. Casalino, M. Theoretical Investigation of Near-Infrared Fabry–Pérot Microcavity Graphene/Silicon Schottky Photodetectors Based on Double Silicon on Insulator Substrates. *Micromachines* **2020**, *11*, 708. [CrossRef] [PubMed]
20. Casalino, M. Design of resonant cavity-enhanced schottky Graphene/silicon photodetectors at 1550 nm. *J. Light. Technol.* **2018**, *36*, 1766–1774. [CrossRef]
21. Echtermeyer, T.; Britnell, L.; Jasnos, P.; Lombardo, A.; Gorbachev, R.; Grigorenko, A.; Geim, A.; Ferrari, A.C.; Novoselov, K. Strong plasmonic enhancement of photovoltage in graphene. *Nat. Commun.* **2011**, *2*, 1–5. [CrossRef] [PubMed]
22. Goykhman, I.; Sassi, U.; Desiatov, B.; Mazurski, N.; Milana, S.; De Fazio, D.; Eiden, A.; Khurgin, J.; Shappir, J.; Levy, U.; et al. On-chip integrated, silicon–graphene plasmonic Schottky photodetector with high responsivity and avalanche photogain. *Nano Lett.* **2016**, *16*, 3005–3013. [CrossRef]
23. Konstantatos, G.; Badioli, M.; Gaudreau, L.; Osmond, J.; Bernechea, M.; De Arquer, F.P.G.; Gatti, F.; Koppens, F.H. Hybrid graphene–quantum dot phototransistors with ultrahigh gain. *Nat. Nanotechnol.* **2012**, *7*, 363–368. [CrossRef] [PubMed]
24. Casalino, M.; Sassi, U.; Goykhman, I.; Eiden, A.; Lidorikis, E.; Milana, S.; De Fazio, D.; Tomarchio, F.; Iodice, M.; Coppola, G.; et al. Vertically illuminated, resonant cavity enhanced, graphene–silicon Schottky photodetectors. *ACS Nano* **2017**, *11*, 10955–10963. [CrossRef]
25. Fowler, R.H. The analysis of photoelectric sensitivity curves for clean metals at various temperatures. *Phys. Rev.* **1931**, *38*, 45. [CrossRef]
26. Elabd, H.; Kosonocky, W.F. PtSi Infrared Schottky-Barrier Detectors With Optical Cavity. *Review* **1982**, *43*, 569.
27. Cohen, J.; Vilms, J.; Archer, R.J. *Investigation of Semiconductor Schottky Barriers for Optical Detection and Cathodic Emission*; Technical Report; Hewlett-Packard Co.: Palo Alto, CA, USA, 1968.
28. Vickers, V.E. Model of Schottky Barrier Hot-Electron-Mode Photodetection. *Appl. Opt.* **1971**, *10*, 2190–2192. [CrossRef] [PubMed]
29. Raissi, F.; Far, M.M. Highly sensitive PtSi/porous Si Schottky detectors. *IEEE Sens. J.* **2002**, *2*, 476–481. [CrossRef]
30. Raissi, F. A possible explanation for high quantum efficiency of PtSi/porous Si Schottky detectors. *IEEE Trans. Electron Devices* **2003**, *50*, 1134–1137. [CrossRef]
31. Geim, A.; Novoselov, K. The rise of graphene. *Nat. Mater.* **2007**, *6*, 183–191. [CrossRef]
32. Nair, R.; Blake, P.; Grigorenko, A.; Novoselov, K.; Booth, T.; Stauber, T.; Peres, N.; Geim, A. Fine structure constant defines visual transparency of graphene. *Science* **2008**, *320*, 1308. [CrossRef] [PubMed]
33. Sze, S.M.; Ng, K.K. *Physics of Semiconductor Devices*; John Wiley & Sons: Hoboken, NJ, USA, 2006.
34. Neto, A.C.; Guinea, F.; Peres, N. KS No voselov, and AK Geim. *Rev. Mod. Phys.* **2009**, *81*, 109.
35. Yu, Y.J.; Zhao, Y.; Ryu, S.; Brus, L.E.; Kim, K.S.; Kim, P. Tuning the graphene work function by electric field effect. *Nano Lett.* **2009**, *9*, 3430–3434. [CrossRef]
36. Takahashi, T.; Tokailin, H.; Sagawa, T. Angle-resolved ultraviolet photoelectron spectroscopy of the unoccupied band structure of graphite. *Phys. Rev. B* **1985**, *32*, 8317–8324. [CrossRef] [PubMed]
37. Ackert, J.; Fiorentino, M.; Logan, D.; Beausoleil, R.; Jessop, P.; Knights, A. Silicon-on-insulator microring resonator defect-based photodetector with 3.5-GHz bandwidth. *J. Nanophot.* **2011**, *5*, 0595071. [CrossRef]
38. Ackert, J.; Knights, A.; Fiorentino, M.; Beausoleil, R.; Jessop, P. Defect enhanced silicon-on-insulator microdisk photodetector. In Proceedings of the Optical Interconnects Conference, Santa Fe, NM, USA, 20–23 May 2012; Volume TuP10, pp. 76–77.
39. Xu, S.; Wang, W.; Huang, Y.C.; Dong, Y.; Masudy-Panah, S.; Wang, H.; Gong, X.; Yeo, Y.C. High-speed photo detection at two-micron- wavelength: Technology enablement by GeSn/Ge multiple-quantum-well photodiode on 300 mm Si substrate. *Opt. Express* **2019**, *27*, 5798. [CrossRef] [PubMed]

Review

Polarization-Sensitive Digital Holographic Imaging for Characterization of Microscopic Samples: Recent Advances and Perspectives

Giuseppe Coppola and Maria Antonietta Ferrara *

National Research Council (CNR), Institute of Applied Sciences and Intelligent Systems, Via Pietro Castellino 111, 80131 Naples, Italy; giuseppe.coppola@cnr.it
* Correspondence: antonella.ferrara@na.isasi.cnr.it

Received: 28 May 2020; Accepted: 22 June 2020; Published: 29 June 2020

Featured Application: A simple and fast measure of the state of polarization of vector optical beams is a very important topic to study new optical effects and their applications in several fields, such as microelectronics, micro-photonics, remote sensing and bioimaging.

Abstract: Polarization-sensitive digital holographic imaging (PS-DHI) is a recent imaging technique based on interference among several polarized optical beams. PS-DHI allows simultaneous quantitative three-dimensional reconstruction and quantitative evaluation of polarization properties of a given sample with micrometer scale resolution. Since this technique is very fast and does not require labels/markers, it finds application in several fields, from biology to microelectronics and micro-photonics. In this paper, a comprehensive review of the state-of-the-art of PS-DHI techniques, the theoretical principles, and important applications are reported.

Keywords: digital holography; polarization sensitive imaging; birefringence; state of polarization (SoP)

1. Introduction

Digital holography (DH) is a fascinating alternative to conventional microscopy since it allows three-dimensional (3D) reconstruction, phase contrast images, and an improved focal depth [1–5]. Basically, DH consists of an interference fringe pattern between a reference unperturbed beam and an object beam, that changes its characteristics by passing through a sample. The interference pattern (hologram) is acquired by a digital sensor array. Its post-processing achieves a 3D quantitative image of the sample by a numerical refocusing of a 2D image at different object planes [6]. When DH is implemented in an optical microscope, the objective lens provides a magnified image allowing to reconstruct amplitude and phase-contrast images with a spatial resolution of less than 1μm in all dimensions [7].

Digital holographic imaging (DHI) has several interesting features including high-resolution, very fast acquisition, and 4D (3D + time) characterization of samples [8–10]. These properties are very useful, for example, when the specimen is moving or when the sample is subjected to external stimuli that can alter its shape and size, such as electrical, magnetic or mechanical forces, chemical corrosion, or evaporation and deposition of further materials. Moreover, DHI is a non-contact and non-invasive technique, allowing label-free quantitative phase analysis of living cells; thus, measurements do not require the introduction of a tag, so cells are not altered. This approach can provide useful information that can be interpreted into many underlying biological processes.

During the last decade, DHI has experienced several technological developments, including the integration of DHI with complementary characterization techniques (e.g., Raman spectroscopy or scanning electron microscope [11–13]). A further important extension of DHI is the possibility to quantitative measure the state of polarization (SoP) modified by a sample [14–17] and so evaluate its birefringent and/or dichroic proprieties, which are frequently related to the micro- or even ultra-structure of the sample itself [18]. Therefore, the characterization of these proprieties and the detection of their eventual variations, that can be due to either stress and strain in a given material or disordered microstructure in biological specimens, could lead to a better understanding of the process involved in a broad variety of applications.

Since SoP is one of the fundamental properties of light, its evaluation has attracted a growing interest in both the basic researches and practical applications of optics, intending to study novel optical phenomena and new applications. Thus, the experimental evaluation of the SoP has become a fast-rising subject. Typically, polarization imaging has been carried out with different approaches—for example by using real-time polarization phase-shifting system [19], polarization contrast with near-field scanning optical microscopy [20], optical coherence tomography [21,22], and Pol-Scope [23]. However, these techniques need different image acquisitions, generally obtained at diverse orientations of birefringent optical components (e.g., polarizers, quarter-wave, and/or half-wave plates) to retrieve the polarization state. The great advantage offered by polarization-sensitive digital holographic imaging (PS-DHI) is the possibility to use a single acquisition to retrieve the full polarization state of the sample under observation, therefore gaining in speed and simplicity.

This review paper aims to provide an overview of the state-of-the-art in PS-DHI. In the following sections, some basic concepts will be introduced for describing polarization of light and commonly used technical approaches for realizing PS-DHI. Then, some recent and important applications of PS-DHIM in both the biomedical field and non-biomedical use will be discussed.

2. Theoretical Background

The Stokes vectors and Müeller matrices allow a whole study of the polarization state for fully polarized, partially polarized and even unpolarized light, comprising the optical axis and the degree of polarization. On the other hand, the Jones vectors, that can be useful only for completely polarized light, are more appropriate for problems concerning coherent light (see Appendix A). As a general rule, the Jones vectors are useful for problems involving amplitude superposition, while the Müeller matrices are applied for problems involving intensity superposition [24]. Different approaches of PS-DHI have been proposed in the literature, however, the basic idea is to generate a hologram of the sample through the interference between the object wave and two orthogonally polarized reference waves, producing in this way two fringe patterns. The hologram of the magnified sample is recorded by a digital camera (such as a charge-coupled device or an active pixel sensor). The numerical reconstruction of such hologram leads to two wavefronts, one relative to each reference wave, and thus, one for each perpendicular state of polarization [15]. Basically, PS-DHI approaches can be classified in two groups—(i) those which allow measurement of Jones vectors or Jones matrices and (ii) those which give information on Stokes vectors or Müller matrices. Since holography needs a uniform laser beam, especially regarding the flatness of phase front and the extended depth of field, in both approaches, (quasi)-monochromic light and perfect plane wavefronts are considered. However, the realistic intensity distribution of laser sources is described by Gaussian function, leading to problems in holographic-based applications, such as a reduced image contrast. These issues can be overcome by implementing beam shaping systems built on the base of field mapping refractive beam shapers like πShaper [25].

2.1. PS-DHI for Jones Formalism

In most polarimetric techniques, SoP parameters can be evaluated by applying more or less complex algorithms to various images acquired with different settings of polarization-analyzing components (polarizers, rotators, and retarders). Since these procedures need several rotations of the analyzing optics, the acquisition time is very long compared to the performances of a digital camera. Several solutions were proposed in the literature to improve the temporal resolution, such as the possibility to use a liquid-crystal universal compensator [23]; however, the goal was reached only with techniques that allow recording all parameters of the polarization state through the acquisition of a single image. Hence, since the pioneering paper published by Ohtsuka and Oka, which generated the interference between two orthogonal linearly polarized reference waves and an object wave [26] using a Mach-Zender interferometer, several other published works followed this approach [15,27,28].

The typical experimental configuration for the recording of polarization holograms is illustrated in Figure 1. It consists of a modified Mach–Zehnder interferometer with two reference waves—R_1 and R_2—that interfere with the object beam O [11,15–17]. Two operating conditions are possible [18]. First, the object plane beam is linearly polarized by a polarizer ($P1$ oriented at 45°) and a quarter wave plate ($QWP1$ oriented at 0° respect the incoming light). Due to the passage through the sample, the state of polarization of the beam O can change. An objective lens is used to collect the wave emitted from the sample. A polarized beam splitter (PBS) allows the user to obtain two orthogonal linearly polarized reference beams R_1 and R_2, where orthogonality avoids any interference between the reference beams. Additionally a couple of polarizers or quarter waves plates ($QWP2$ and $QWP3$) preserve the linear polarization when their fast axes are aligned parallel respect to the polarization states of the respective reference waves. In the second operating condition, a circularly polarized light is incident on the sample by orienting the $QWP1$ fast axis angle at −45° (left-handed circular polarization) or +45° (right-handed circular polarization) respect to the polarizer transmission axis $P1$. The two orthogonal linearly polarized reference are transformed in right and left circularly polarized beams by the quarter wave plates $QWP2$ and $QWP3$ oriented of +45° and −45° respect to the polarization states of the two references waves, respectively. The beam splitter $BS1$ allows to overlap O, R_1, and R_2 beams, and the interference among these waves gives rise to the hologram that is acquired by a digital camera in an off-axis configuration, i.e., with the three waves propagating along slightly different directions (as highlighted in the inset in Figure 1). The intensities of O, R_1 and R_2 can be adjusted by the half-waves plates $HWP1$ and $HWP2$, while the angles of incidence of R_1 and R_2 can be controlled by mirrors $M2$ and $M3$, respectively. An example of the recorded digital hologram in shown in Figure 2a. In the insert, where a magnification of the interference between object and the two orthogonal reference waves is reported, the two sets of fringe patterns are clearly visible.

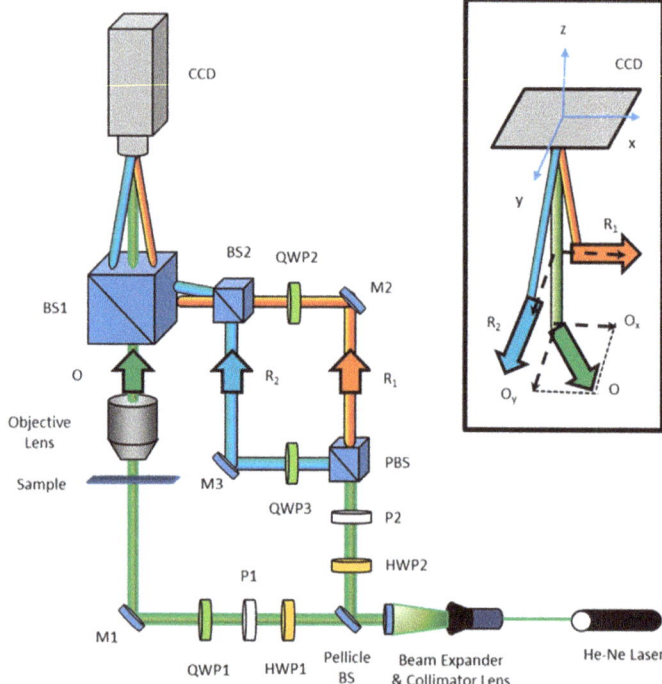

Figure 1. Basic scheme of the polarization-sensitive digital holographic imaging (PS-DHI) experimental setup. Abbreviations: M: mirrors; BS: beam splitter; PBS: polarized beam splitter; P: polarizer; QWP: quarter-wave plate; HWP: half-wave plate. In the inset, the incident directions and the polarization states of the object and reference waves are highlighted.

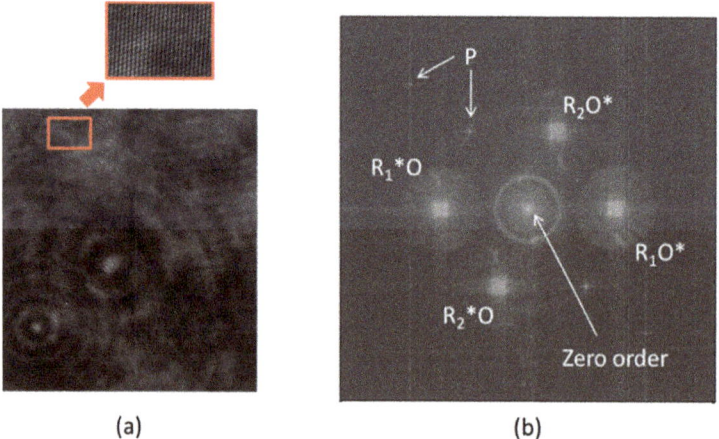

Figure 2. (a) Example of polarization hologram; in the inset the two fringe patterns are highlighted; (b) Fourier amplitude spectra of the hologram: the frequencies of the zero-order of diffraction, of the virtual and real images, and of parasitic interferences (P) are clearly visible.

Since R_1 and R_2 are orthogonally polarized, they do not interfere: $R_1 \cdot R_2^* = R_1^* \cdot R_2 = 0$ (where the asterisk indicates the complex conjugate), so the hologram intensity at the digital camera surface is [11,15]

$$\begin{aligned} H(x,y) &= (O + R_1 + R_2) \cdot (O + R_1 + R_2)^* \\ &= |O|^2 + |R_1|^2 + |R_2|^2 + OR_1^* + OR_2^* + O^*R_1 + O^*R_2 \end{aligned} \quad (1)$$

In Equation (1), the first three terms correspond to the zero diffraction order, the fourth and fifth terms produce the virtual images, and the last two terms form the real images. Computing the Fourier transform of the acquired hologram, these terms appear spatially separated in the Fourier space, due to the off-axis configuration, as well shown in Figure 2b where the presence of some parasitic interferences are also highlighted [16,29]. In order to recover the information of the real images, the corresponding spectra could be selected by two different spatial filters [15,18].

As with classical holography, the reconstruction is obtained by multiply the digital hologram for a digitally computed reference wave and then the inverse Fourier transform of the spatial frequency components filtered is performed. By using the standard reconstruction algorithm on each filtered region, the two orthogonal components of the object beam (O_x and O_y), can be retrieved [15,16,18,30]. So, the amplitude map and the phase map for each polarization component can be reconstructed. With the aim to evaluate the SoP change of the object beam due to the interaction with the sample under test, typically, two parameters are experimentally measured—the amplitude ratio β, which is related to the different transmitted intensities of the two orthogonal components and corresponds to the azimuth of the polarization ellipse, and the phase difference $\Delta\varphi$, that contains information on the different optical paths due to the refractive index anisotropy linked to the sample structure [18,30].

So, the amplitude ratio angle can be evaluated as

$$\beta = \arctan\left(\frac{|O_y|}{|O_x|}\right) \quad (2)$$

Equation (2) is obtained assuming that both the reference waves have the same intensity ($|R_1| = |R_2|$); this identity is experimentally achieved by controlling the orientation of the half-wave plates *HWP1* and *HWP2* in Figure 1 [18,30]. Regarding the phase difference between the orthogonal components of the object beam, it can be expressed as

$$\Delta\varphi = phase(O_y) - phase(O_x) + \Delta\varphi_R \quad (3)$$

where $\Delta\varphi_R = phase(R_2) - phase(R_1)$ can be removed by a calibrated phase difference offset superimposed to the phase difference image [31]. The evaluation of β and $\Delta\varphi$ allows the SoP of the sample under test to be univocally obtained; in other words, the distributions of the Jones vector at the surface of the sample under test can be retrieved from a single hologram acquisition [31]. For example, if $\Delta\varphi = 0$ or π a linear polarization is retrieved, while if $\Delta\varphi = \pi/2$ and $\beta = \pi/4$ or $\Delta\varphi = -\pi/2$ and $\beta = \pi/4$ a circular right or left polarization, respectively, are detected. Elliptical polarization states, i.e., intermediate polarization values, can be also measured. Considering the polarization ellipse represented in Figure 3 and that corresponds to the projection of the trajectory of the extremity of the vector *O* on the plane xy, it can be characterized by the parameters ψ (orientation angle) and χ (ellipticity angle), which can be additionally evaluated by the following equations [15,17]:

$$\begin{aligned} \psi &= \tfrac{1}{2}\arctan\left(\frac{2|O_x||O_y|\cos(\Delta\varphi)}{|O_x|^2 - |O_y|^2}\right) = \tfrac{1}{2}\arctan[\tan(2\beta)\cos(\Delta\varphi)] \\ \chi &= \tfrac{1}{2}\arcsin\left(\frac{2|O_x||O_y|\sin(\Delta\varphi)}{|O_x|^2 + |O_y|^2}\right) = \tfrac{1}{2}\arcsin[\sin(2\beta)\sin(\Delta\varphi)] \end{aligned} \quad (4)$$

Appl. Sci. 2020, 10, 4520

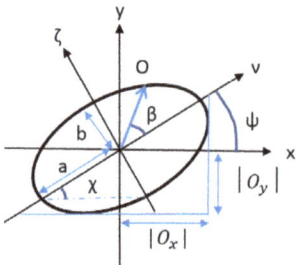

Figure 3. Polarization ellipse. The ellipticity is given by the ratio of the length of the semiminor axis to the length of the semimajor axis, b/a = tan(χ). The ellipse is further described by its azimuth ψ, measured counterclockwise from the x axis [10].

Regarding the direction of rotation of the vector O, it is defined as left-hand polarization (L-state) or right-hand polarization (R-state) depending on whether $-\pi \leq \Delta\varphi \leq 0$ or $0 \leq \Delta\varphi \leq \pi$, respectively.

Finally, in a recent and very interestingly work a new compact experimental setup has been proposed as proof of concept [32]. The birefringence distribution was measured by the interference fringes based on three circularly polarized beams—two mutually orthogonal polarized (left-handed and right-handed) reference waves and a right-handed object wave. These three-beam interfering fields were obtained by monolithic gratings, so all waves were crossed at the same angle on the hologram plane, generating two sets of the fringe pattern. With this approach, all the optical elements required to obtain the three beams in the set-up illustrated in Figure 1 can be replaced by three grating vectors; moreover, in order to have practical interferometry, a monolithic grating containing three diffractive gratings positioned in a threefold-symmetric arrangement has been employed [32,33]. A schematization of both the monolithic gratings and the operating principle is illustrated in Figure 4. When a plane beam impinges on the monolithic gratings, the first-order diffractions of each grating intersect the hologram plane with the same angles. As described, analyzing the interference patterns allows the estimation of the anisotropic phase shift in the object beam. Thus, this arrangement has the potentiality to evaluate a two-dimensional birefringence distribution in a single shot and compact way.

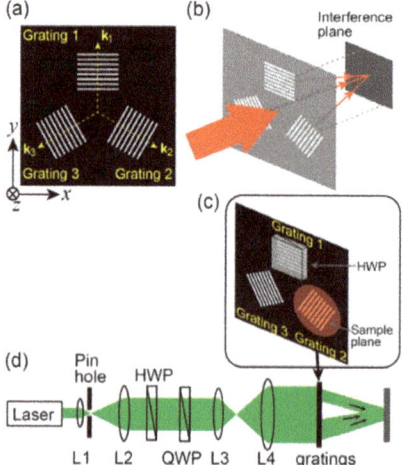

Figure 4. Representation of (**a**) the monolithic gratings, (**b**) interference of the diffracted beams, (**c**) monolithic gratings with the half-wave plate (HWP) and sample of birefringent medium, and (**d**) experimental setup designed for measuring birefringence distributions (Reproduced with permission from Shimomura et al. [32], © The Optical Society, 2018).

2.2. PS-DHI for Stokes Vectors or Müller Formalism

Jones vectors cannot be used to describe light that does not remain in a single polarization state. So, to treat fully, partially, or unpolarized light, Stokes parameters (S_1, S_2, S_3) are often used. In fact, unlike the Jones vectors, the Stokes parameters can define all kinds of SoPs of the optical beam. Unfortunately, as in the case of Jones formalism, the typical methods for estimating the Stokes parameters of an optical wave require to register several intensity distributions at different detection states via rotating the polarized optical elements. Generally, these methods require time-sequential operations using an arrangement with a rotating waveplate and a fixed analyzer. However, system error could be generated by the rotating elements, due to their inhomogeneous transmittance, and obviously, fast acquisitions are not possible [34,35]. Other proposed methods need multichannel simultaneous measurement; in this case, the amplitude or the wavefront is split into several channels, each one is analyzed by employing appropriate polarization optical components thus leading to a complicated and cumbersome dynamic measurement system [36,37]. Moreover, systematic and calculation errors are induced by a not exactly precise image matching.

Thus, it is clear that the development of imaging polarimeters in real-time without the demand for mechanical or active elements for polarization control is still a valuable aim. With this purpose, several approaches were proposed; in one of these, multiple interference patterns are produced at the surface of a digital camera and the information of a different polarized component of the optical beam under investigation is linked to fringes patterns having different spatial frequencies. The Stokes parameters can be estimated through demodulation of the obtained image by a Fourier transform approach [38]. Recently, the combination of DH with theory of the Pancharatnam–Berry (PB) phase has paved the way toward the implementation of a new method for evaluating the state of polarization of arbitrary waves with a single exposure of the interference pattern and a quick acquisition for one object wave with no moving optical components [39,40]. Pancharatnam [41] and Berry [42] introduced the so-called PB phase of an optical wave that is taken along a closed cycle on the Poincaré sphere. With this formalism, polarization transformations give rise to two optical phase retardation—one related to optical path difference (called dynamic phase) and an extra one which is equal to minus half of the solid angle subtended by the closed path on the sphere. Therefore, this extra phase, i.e., the PB phase, depending only on the geometry of the transformations' path of on the Poincaré sphere, is also called "geometric phase." Nevertheless, the PB phase occurs generally in the following two conditions: when there is a variation of polarization state in the beam propagation, and when there is a variation of the mode structure of the beam propagation [43,44]. Regarding the first condition, the PB phase and the polarization state variation are quantitatively related, giving the possibility to estimate the SoP from the PB phase measurement. Even though DH allows quantitative phase retrieve of the object beam to be performed [45,46], when the phase difference contains both the PB phase (related to the SoP) and dynamic phase (related to the optical density of the sample), these two phases are indistinguishable for the reconstruction process. To separate the two contributions, two holograms should be generated—one to retrieve the dynamic phase and the other one for the geometric phase, respectively. The dynamic phase is then used for the evaluation of the refractive index or 3D shape of the sample under investigation, whereas its interaction with the polarized light is estimated through the geometric phase.

Basically, two classes of experimental setup based on the geometric phase can be found in the recent literature to evaluate the full SoP of an optical beam. In the first, shown in Figure 5a, a triangular common-path interferometer (TCPI) is used to generate two interferograms which are aligned together on a single charge coupled device (CCD) target [40]. This is made possible by dividing the object beam into two orthogonal circularly polarized components (left-handed and right-handed) through the TCPI, and then these two components interfere with a reference beam. However, this implementation is difficult to align, leading to low measurement resolution, reduced field of view (due to separated fringe pattern for the orthogonal components recorded in two different region of the same CCD), and image matching errors. The second setup, reported in Figure 5b, is based on a hybrid polarization-angular multiplexing digital holographic approach (PAMDH) [47], implemented by a double-channels Mach–Zehnder interferometer. In this case, two orthogonal and linearly polarized reference beams interfere with the object beam.

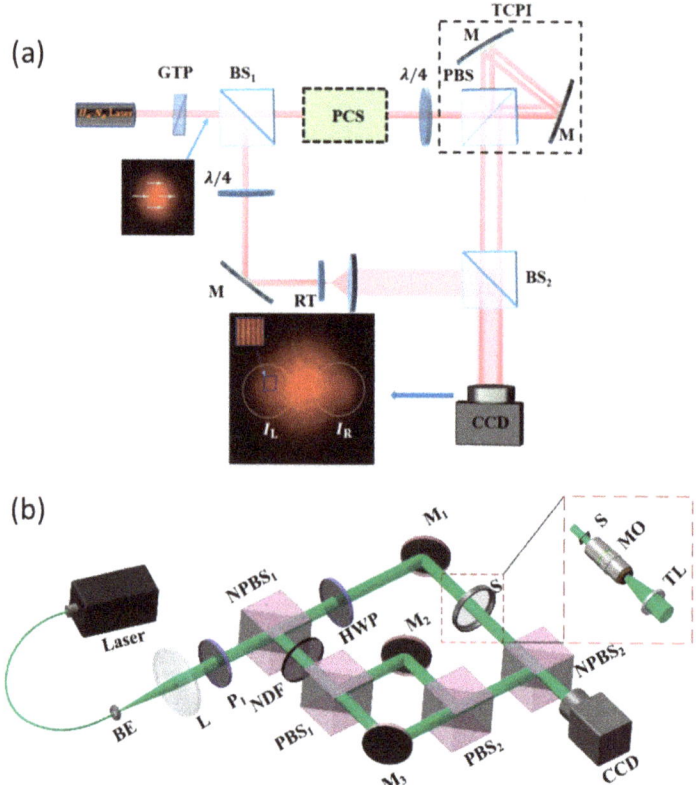

Figure 5. Experimental setup for polarization measurement—(**a**) triangular common-path interferometer (TCPI). GTP: Glan–Taylor polarizer; BS1 and BS2: beam splitter; PCS: polarization conversion system; λ/4: quarter-wave plate; PBS: polarizing beam splitter; M: mirror; RT: reversed telescope; CCD: charge-coupled device. The elements inside the dashed box form a TCPI (Reproduced with permission from Qi et al. [40], AIP, 2019); (**b**) schematic of the PAMDH system based on the geometric phase. BE: beam expander; L: lens; P: polarizer; S: sample; HWP: half-wave plate; NPBS$_1$-NPBS$_2$: non-polarized beam splitters; PBS$_1$-PBS$_2$: polarized beam splitters; M$_1$-M$_3$: mirrors; NDF: neutral density filter; MO: microscope objective; TL: tube lens (reprinted from Dou et al. [47]).

In both cases, the acquisition of a composite hologram generated by combining two patterns of interference fringes with distinct orientations allows the contemporary estimation of the orthogonal polarization components for an optical wave. Once these components are retrieved, the Stokes parameters can be evaluated by applying the geometric phase theory [47]. The main difference between two reported schemes is that in the setup depicted in Figure 5b the multiplexed hologram covers the whole area of the CCD, while in the solution presented in Figure 5a, a smaller field of view is achieved to avoid the influence of the change in the intensity and polarization distribution of the reference wave. For this configuration, the field of view is determined by the region of two images on CCD related to the two orthogonal components of the object wave separated at a distance controlled by the TCPI.

Therefore, since the arrangement of Figure 5b is more useful to produce and control a larger field of view, for simplicity of discussion, here, the basic theory related to this scheme is reported.

In the PAMDH approach, Equation (1) becomes [47]

$$\begin{aligned} H(x,y) &= (O + R_V + R_H) \cdot (O + R_V + R_H)^* \\ &= |O_V|^2 + |O_H|^2 + |R_V|^2 + |R_H|^2 + O_V R_V^* + O_H R_H^* + O_V^* R_V + O_H^* R_H \\ &= I_V + I_H \end{aligned} \quad (5)$$

$$\begin{aligned} I_V &= |O_V|^2 + |R_V|^2 + 2|O_V||R_V|\cos\varphi_V \\ I_H &= |O_H|^2 + |R_H|^2 + 2|O_H||R_H|\cos\varphi_H \end{aligned} \quad (6)$$

where φ_V and φ_H are the phase variations related to each of the two orthogonal components of the reference beams and object waves. Therefore, the two orthogonal complex amplitudes O_V and O_H of the object wave can be numerically retrieved. To reconstruct the SoP, the polarization state should be reported on the Poincaré sphere. With this aim, in Figure 6, a spherical coordinate system is shown; here, the vertical and horizontal states (V,H) are positioned on the two poles, whereas the polar and azimuthal angles are given by $2\chi_1$ and $2\psi_1$, respectively.

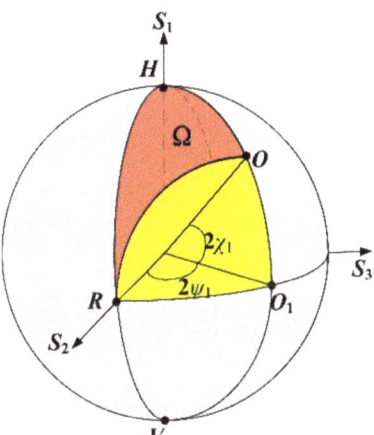

Figure 6. Theoretical model for estimating polarization state based on geometric phase (reprinted from Dou et al. [47]).

In accord with Pancharatnam's theory [41], the orthogonal components intensities of the object wave I_O are given by

$$\begin{aligned} I_{OV} &= |O_V|^2 = I_O \sin^2(\hat{OH}/2) \\ I_{OH} &= |O_H|^2 = I_O \cos^2(\hat{OH}/2) \end{aligned} \quad (7)$$

Once the distribution of the two components of the complex amplitude of the object wave has been retrieved, the polar angle can be evaluated as

$$2\chi_1 = \pi/2 - \hat{OH} = \pi/2 - 2\arctan|O_V|/|O_H| \quad (8)$$

As previously discussed, for each polarization state V and H, the phase difference between reference and the object waves contains the contributes of both the dynamic and the geometric phase, in particular,

$$\varphi_V = \varphi_{Vd} + \varphi_{Vpb} \quad (9)$$

$$\varphi_H = \varphi_{Hd} + \varphi_{Hpb} \quad (10)$$

In Equations (9) and (10), the first term in the sum denotes the dynamic phase difference and the second term is the PB phase. For horizontal polarization state, the PB phase is numerically equivalent to half of the area of the geodesic triangle ROH (the red region in Figure 6), i.e., $\varphi_{Hpb} = \Omega/2$ [48]. Correspondingly, for vertical polarization state, the PB phase is given by $\varphi_{Vpb} = -2\psi_1 + \Omega/2$. So, the azimuthal angle can be evaluated as [47]

$$2\psi_1 = (\varphi_H - \varphi_V) - (\varphi_{Hd} - \varphi_{Vd}) \tag{11}$$

Taking advantage by DH, the two orthogonal components of the optical field can be obtained by

$$\begin{aligned} E_V &= k|O_V||R_V|e^{i\varphi_V} \\ E_H &= k|O_H||R_H|e^{i\varphi_H} \end{aligned} \tag{12}$$

where k is a constant depending on the exposure time and response of the digital camera. The holographic reconstruction approach allows evaluating the phases φ_V and φ_H, as a consequence the first term in brackets on the right side of Equation (11) can be calculated. Regarding the second term in brackets, $\Delta\varphi = (\varphi_{Hd} - \varphi_{Vd})$, it is a fixed value for the adjusted PAMDH configuration and can be corrected by measuring a linearly polarized beam along 45° produced by a standard polarizer.

Since the two reference waves are adjusted to have the same intensity, the azimuthal and polar angles can be estimated by the following relationships:

$$\begin{aligned} 2\psi_1 &= arg(E_H/E_V) - \Delta\varphi \\ 2\chi_1 &= \pi/2 - 2arctan(|E_V|/|E_H|) \end{aligned} \tag{13}$$

Thus, the area of Ω can be evaluated as a function of the angles $2\psi_1$ and $2\chi_1$ [39] and the dynamic phases φ_{Hd} and φ_{Vd} are calculated [40,47–49]. Finally, the normalized Stokes parameters (S_1, S_2, S_3) can be expressed as a function of ($2\psi_1, 2\chi_1$) with a trigonometric relationship, as shown in Figure 6 and as described by the following relations [44]:

$$\begin{aligned} S_1 &= \sin 2\chi_1 \\ S_2 &= \cos 2\chi_1 \cos 2\psi_1 \\ S_3 &= \cos 2\chi_1 \sin 2\psi_1 \end{aligned} \tag{14}$$

The so-evaluated Stokes parameters describe the state of polarization of the object wave at the measurement area, hence they fully characterize the SoP of an arbitrarily polarized wave.

Finally, it is worth noting that the signal light from some imaging techniques, such as fluorescence imaging, is spatially incoherent, and therefore requires a further challenge for holographic imaging. Moreover, the use of light sources such as lasers, with highly temporal and spatial coherence, reduces the quality of the hologram owing to speckle and spurious fringe generation, which can decrease the spatial sensitivity of the system [50,51]. Consequently, polarization-sensitive imaging of an incoherent scene or by using partially incoherent illumination should be implemented. Unfortunately, only a couple of work proposed the incoherent polarization sensitive holography/interferometry. For example, Zhu and Shi [52] investigated a self-interference polarization holographic imaging (Si-Phi) method that allows real-time 3D imaging of an incoherent scene. The authors developed an in-line polarization holography configuration equipped with a polarization-resolving detector array; this setup allows a single shot acquisition of the complex-valued hologram and results demonstrated both 3D and real-time imaging capabilities. Even if the use of incoherent sources is still immature, future developments are expected in this field.

3. PS-DHI Applications

As described in the previous section, PS-DHI can measure the parameters β and $\Delta\varphi$, thus allows to retrieve the SoP of a beam that interacts with a specimen. The modification of the SoP in the transmitted or reflected beam gives information about the structure, the composition, or the optical properties of the specimen under study. Basically, the following two physical properties of the matter can alter the polarization state of a wave [18]:

- Birefringence—a material is considered birefringent if its refractive index depends on the polarization and propagation direction of the incoming light, i.e., it shows an optical anisotropy. When these samples are crossed by a polarized light, the amplitudes are unchanged but a modification in the relative phase occurs. Birefringence can be linear (that is, there is one axis of symmetry, called the optic axis) such as in optical wave plates/retarders and many crystals, or circular (that is, in which for an incident linearly polarized light, the corresponding outgoing polarization plane will be rotated) such as chiral fluids.
- Dichroism—a material, typically crystalline, is considered dichroic if it absorbs more light along a preferential incident plane of polarization than another plane (absorption anisotropy); as a result, when the optical beam propagates within this material, its polarization state undergoes a modification. The ratio of amplitudes of the orthogonal components of the light emerging from the sample under test provides a measurement of its linear dichroism property.

The study of the polarization state covers different applications and research fields, such as measurement of stress, geology, chemistry, display technologies, medicine and medical diagnosis, etc. Currently, it has been demonstrated that the PS-DHI technique can be used for noninvasive quantitative imaging of live cells or the evaluation of the dynamic phase difference induced by the birefringence of liquid crystals. In the following, a state of the art of the PS-DHI applications is reported in two subsections, dividing the biological from microelectronics and micro photonics applications.

3.1. Microelectronics and Nanophotonic Quantitative Phase Imaging

Since PS-DHI has been introduced in the literature, it has been applied on samples with well-known SoP, such as bent fiber [16], stressed polymer [15], waveplates [39,40], or liquid crystals (LCs) [47], just to confirm the potentiality offered by this technique. Among these applications, LC seems to be the most interestingly due to their uses in the display. For example, Park et al. [49] measured the spatially resolved Jones matrix components of the light passing through the single pixel of a liquid-crystal display (LCD) as a function of the applied voltage to the LCD panel by using PS-DHI. However, in the proposed setup, the authors need to acquire four independent interferograms with two different polarization states to reconstruct Jones matrix components map from a sample, thus the properties of PS-DHI are not fully exploited. In the published work, an in-plane switching liquid-crystal display (IPS-LCD) was characterized. However, the same approach can be used to feature also other types of LCDs—for example, full RGB channels of LCD pixels can be characterized in terms of their Jones matrix components using a DH setup with multiple lasers or in spectroscopic modality.

An LC depolarizer was characterized using PS-DHI with the scheme reported in Figure 5b by Dou et al. [47]. In this case, the LC depolarizer consists in a collection of HWPs with optical axes randomly distributed and the Stokes parameters distribution of the output beams for a linear incident polarized beam with $\theta = \pi/4$ and for left-handed circularly polarized incident beam were measured, confirming the depolarizing effect induced by LC [47].

Regarding nanophotonics applications, images obtained with DHI have the disadvantage of being limited by the diffraction limit and, thus, a device in the nanometres size typically covers just a few camera pixels leading to a low resolution. On the other hand, a full field radiation pattern (i.e., polarization, amplitude, and phase) measurement at all angles gives a complete polarizability tomography of nanophotonic devices such as metasurfaces and nanoantennae. For this reason, Röhrich et al. [53] have combined Fourier microscopy, polarimetry, and digital holography, generating a signal over an entire CCD chip, for angle-resolved amplitude, polarization, and phase imaging of single nano-objects. In particular, the authors analyzed the orbital angular momentum (OAM) content of light scattered by a family of plasmonic spirals. In Figure 7 results obtained in Ref. [53] are summarized; the intensity radiation distribution is recorded by a Fourier microscope (Figure 7a—logarithmic scale). Then, the Stokes parameters are determined by polarization-resolved imaging; these parameters completely define the SoP of the wavefront for each wave vector in the radiation pattern and, consequently, can be transformed in the polarization ellipse parameters—namely, the ellipticity and the orientation angles (Appendix A, Equation (A5)). The evaluated ellipse parameters are shown in Figure 7b and a complete helicity conversion in a doughnut-like pattern with five spiraling arms around it can be observed. Finally, by using DHI the individual phase profiles of two orthogonally polarized field components were retrieved; in Figure 7c the representation in the Fourier transform domain of a hologram corresponding to an m = −5 spiral in circular co-polarization is reported, while in Figure 7d the two evaluated phase maps related to the co- and cross-polarized channel are shown and for this latter channel, the helical shape around the optical axis is clearly visible. As suggested by Röhrich et al. [53], this approach can be applied to several nanophotonics problems such as plasmonic oligomer antennae for emission and sensing, metasurfaces for monitoring wavefronts (transmitted and reflected) as a function of the incident amplitude, phase and vector contents, and nonlinear metasurfaces whose efficiency and angular distribution depend on the phase gradients structured in the metasurface geometry.

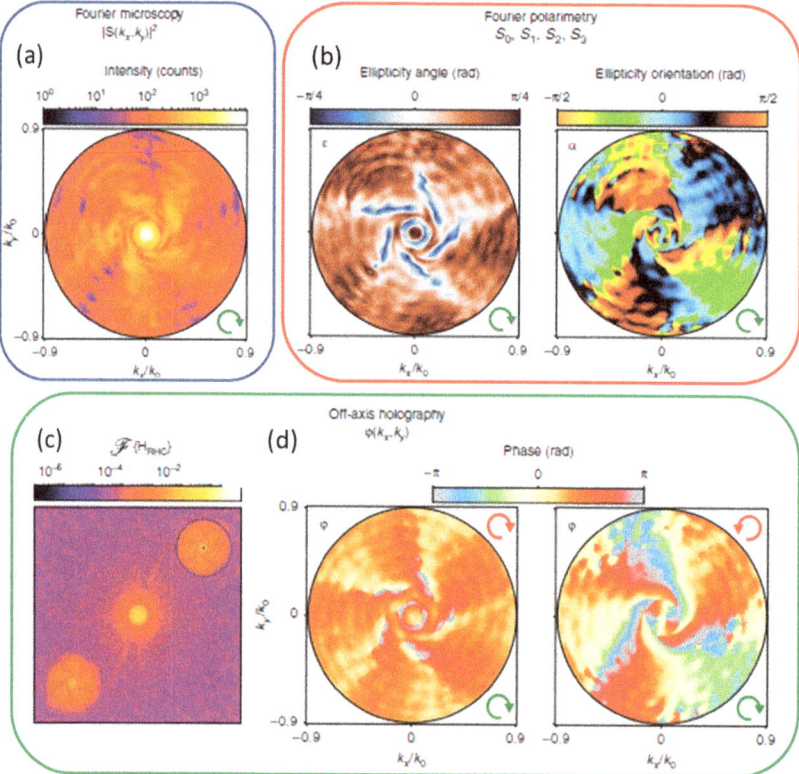

Figure 7. Demonstration of combined Fourier microscopy, polarimetry, and digital holography. R-state circular polarized input and an m = −5 spiral nanostructure were used. (**a**) Fourier map of intensity; (**b**) reconstructed polarization ellipse parameters; (**c**) digital Fourier transform of an interferogram obtained with R-state circular polarized detection; (**d**) reconstructed phase maps for R-state and L-state circular polarized detection. The green and red arrows specify the input and output polarization, respectively (reprinted from Röhrich et al. [53]).

3.2. Biological

The order of molecular architecture can play a role in the dependence of the refractive index on the polarization and propagation direction of the optical beam. Indeed, the anisotropy of the refractive index, i.e., birefringence, could be related to the presence of filament arrays and/or membranes (made of a lipid bilayer that exhibits some degree of orientation) included in organelles and cells. Several pathological modifications, due, for example, to physical damage or disease such as cancer, may modify optical properties of biological tissues by altering their structure and, thus, leading to a change in their birefringence pattern [54–56]. Therefore, the detection of these modifications could become a valuable tool to identify the molecular order, follow events, and diagnose diseases. Typically, the characterization of birefringence is carried out by using quantitative polarized light microscopy [57–59], polarimetry [60–63], and experimental determination of the Müeller matrix [64–66]. Since these techniques require the acquisition of several images to retrieve the birefringence of the sample, they appear too slow for living semi-transparent biological sample imaging.

For this reason, PS-DHI may have a good chance of being used as a label-free and fast technique (only one acquired image) for the study of the SoP of biological samples. However, this technique has started to be used for biological imaging only in the last few years; therefore, the scientific papers on this field of application present in the literature are still a low number. For example, Wang et al. [67] studied the birefringence distribution of biological tissues by using polarization-dependent phase-shifted holograms. Even if the authors used a setup based on a modified Mach–Zehnder interferometer, which allows the recording polarization holograms by rotating a polarizer and so requiring multiple acquisitions, this approach takes advantage by DH and, thus, can be exploited to have only one image acquisition. Interestingly, the results demonstrated that the median birefringence value of cancerous bladder tissues is higher than that of the normal bladder tissues. Hence, this approach can be effectively used to discriminate between cancerous and non-cancerous tissues.

PS-DHI in a configuration similar to that shown in Figure 1 has also been used to distinguish among three different B-leukemia transformed cell lines, providing a diagnosis method of acute lymphoblastic leukemia type B, a cancer with a high mortality rate that affects B lymphocytes [68]. The same approach has been applied also to human sperm cells [30]. In fact, in normal morphological human sperm cells heads, due to longitudinally oriented protein filaments, there is a strong birefringence [69]. When the acrosome reaction occurs, i.e., spermatozoa are ready to approach the egg, the local protein organization disaggregates and, as a consequence, there is a variation in the intrinsic birefringence properties. In fact, reacted spermatozoa show a partial head birefringence, typically in the post-acrosomal region [70]. In the paper published by De Angelis et al. [30], PS-DHI is proposed in a configuration combined with Raman spectroscopy (RS) for a complete, accurate and label-free estimation of the biological proprieties of fixed air-dried sperm. Indeed, PS-DHI provides quantitative information on the cell morphology, motility and SoP [8,71], whereas the RS technique gives complementary specific biochemical fingerprint of the sample, without harming the integrity of live specimens [72]. In Figure 8a,b, the amplitude parameter β and phase difference $\Delta\varphi$ relative to a control sperm cell and to reacted sample retrieved by PS-DHI are reported. It is evident that $\Delta\varphi$ map shows a birefringence distribution (bright pattern) over all the sperm head in the control sperm cell, while the $\Delta\varphi$ map for reacted sperm cell presents a reduction of the birefringent distribution, that is confined in the post-acrosomal area. The acrosome reaction was induced by a heparin treatment. A statistical analysis of the distribution of birefringence patterns of sperm from three donors exposed to heparin for 0 h (control sample) and 4 h (reacted sample) was performed, and the results are resumed in Figure 8c. By combining PS-DHI and RS, the authors proposed a new fully label-free protocol for the recognition of healthy and reacted sperm cells. In detail, sperm cells with a head entirely birefringent are selected by PS-DHI, assuring their integrity; then, the heparin treatment was performed on these chosen spermatozoa to induce the acrosome reaction. Finally, the effectively reacted spermatozoa are selected by estimating again their polarization state by the PS-DHI combined with the study of their Raman spectra. This interesting combined approach leads to the identification of spermatozoa in which the modification in their birefringence distribution is imputed only to the acrosome reaction, while those in which this variation is correlated with defects are not considered [30].

In 2019, Gordon et al. [73] proposed a proof of concept of a holographic fiberscope that allows producing full-field images of amplitude and quantitative phase in two polarizations, using a novel parallelized transmission matrix approach. The polarimetric imaging of birefringent and deattenuating samples was carried out to verify the feasibility of this approach. Due to their small diameter and high flexibility, imaging through optical fibers is already implemented in biomedical endoscopy and industrial inspection. Therefore, the introduction of a holographic fiberscope for birefringence measures appears very interesting for biological applications and remote sensing.

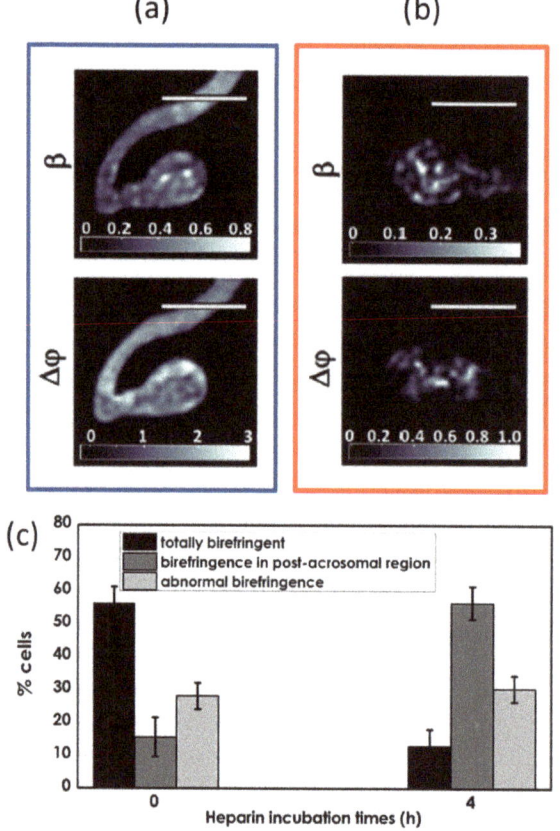

Figure 8. (a) Amplitude parameter β and phase difference $\Delta\varphi$ relative to a control sperm cell (0 h in heparin). Colorbars indicate the mapping of the phase variation. (b) The same polarization parameters measured for a reacted sample (4 h in heparin). (c) Distribution of birefringence patterns of sperm from three donors exposed to heparin for 0 h (control sample) and 4 h (reacted sample). Scale bar: 4 μm (adapted from De Angelis at al. [30]).

An application worthy of note is that studied by Öhman et al. [74,75]. They developed a dual-view polarization-resolved digital holographic systems for particle tracking [74] as well as for define if a particle has a spherical shape or not and to estimate its size [75]. This novel system allows to image the same volume from two perpendicular directions, giving information about the amplitude ratio angle β from two views. The authors found that the size of a non-spherical particle can be estimated from β with an upper limit of about nm. This approach could be very useful to detect and distinguish different particles, including biological particles, under different flow conditions by estimating their polarization response.

Finally, the instantaneous (single-shot) measure of the spatial variations of the phase retardance induced in either geometric or dynamic phase is carried out through an alternative approach presented in Ref [76], where a quantitative fourth-generation optics microscopy (Q4GOM) has been developed; even though this approach doesn't characterize the full SoP, thanks to its unique optical performance, it can open new research in diagnosis of optical composite nanostructures or biomolecular sensing. The phase restoration is based on the self-interference of optical wave and is achieved in an intrinsically stable common-path setup. Basically, in the live cell imaging an add-on fourth-generation (4G) optics imaging module is combined to a polarization adapted interference microscope, as shown in Figure 9a. A light linearly polarized in the same direction of the azimuth of the compensating quarter-wave plate QWP_1 and obtained by the input polarizer P_1, enters in a Mirau microscope objective (MMO) by the beam splitter cube BS_2. Therefore, the object (reflected from the sample) and reference (reflected from the reference mirror M) beams are orthogonally linear polarized after passing twice through QWP_1 and QWP_2 (see Figure 9b). Another quarter-wave plate (QWP_3) transforms the orthogonal linear polarizations into L-state and R-sate circular polarizations. The light collimated by the MMO is focused by the tube lens TL, whose back focal plane corresponds to the input plane of the add-on 4G optics module. Here the L-state and R-sate circular polarizations images created in the sample and reference path overlap, and a polarization directed geometric-phase grating (GPG), with a spatial period Λ = 9 μm (corresponding to 2π rotation of the anisotropy axis) is positioned. The polarization state of the object and reference beams is changed from L-state to R-state circular polarization and vice versa by passing through the GPG, whereas the geometric phase changes as ±2φ, where φ is the periodic spatial change of the angular orientation of the anisotropy axis (see Figure 9c). This geometric-phase modulation leads to a tilt of the object and reference waves with the orthogonal circular polarizations in directions of +1st and −1st diffraction order with the mutual angle of 8° for the central wavelength. The polarizer P_2 and the lens L_2 give the polarization projection and the Fourier transform, respectively; then, the off-axis hologram is recorded on the CCD. By using an optical path difference compensator to the back focal plane of the lens L_1, the length of the object and reference beams optical paths can be aligned, allowing the successfully use of the MMO in biological experiments using broadband light. The Q4GOM has been tested for quantitative imaging of diverse cells classes: human cheek cells, blood smear and spontaneously transformed rat embryonic fibroblast cells. As example, the images obtained for human cheek cells are reported in Figure 9d. Results are very impressive, since they demonstrated an accuracy well below 5 nm, opening new research directions in the quantitative retardance imaging of anisotropic biological samples [76].

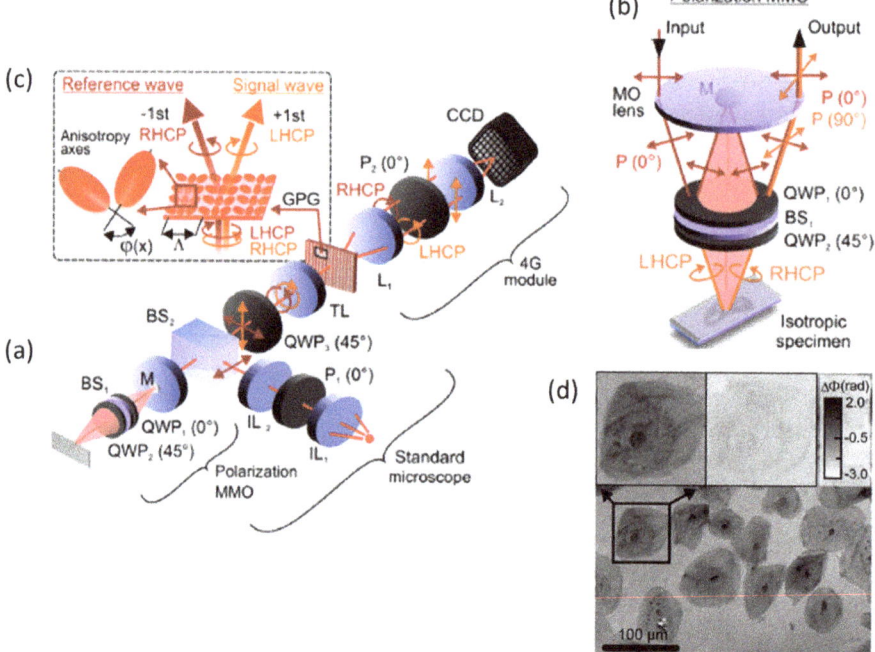

Figure 9. Illustration of quantitative fourth-generation optics microscopy (Q4GOM). (**a**) Experimental setup using 4G optics module connected to microscope with a polarization adapted interference objective. P_1: input polarizer; IL_1, IL_2: illumination lenses; MMO: Mirau microscope objective; BS_1: pellicle beam splitter; QWP_1, QWP_2, QWP_3: quarter-wave plates; M: reference mirror; BS_2: beam splitter cube; TL: tube lens; GPG: geometric-phase grating; L_1: first Fourier lens; P_2: analyzer; L_2: second Fourier lens; CCD: charged coupled device. (**b**) Polarization-adapted Mirau microscope objective (MMO) used for imaging of isotropic samples. (**c**) Polarization sensitive transformation of light by geometric phase grating. (**d**) The quantitative phase retardance imaging of human cheek cells. At the top a comparison of the quantitative phase imaging (left) and bright field image (right) of the marked area (adapted from Bouchal et al. [76]).

4. Conclusions

PS-DHI is a flexible, useful development of DHI; indeed, only a few changes to the standard DH setup are required to obtain polarization-based imaging. However, innovative solutions were also developed. The requirement of a single-shot imaging and high processing speed significantly improve the operation of the measurement process making this approach more appropriate for real-time multiple analyses. Moreover, the full SoP and phase distribution for an arbitrary light field can be easily and quickly measured by PS-DHI based on the geometric phase.

In this context, this review paper presents a brief introduction to the basic principles underlying PS-DHI and an overview of some enhancements in its technology development. To the best of our knowledge, there are no other reviews on this topic. Therefore, it is our belief that this work could help researchers who work in this field. Even if PS-DHI is a fairly established research line (the first work proving its feasibility dates back to 1999 [17], while in the past 20 years, many works have been published to introduce improvements to the technique), there are currently only a few applications presented in the literature. Among these, the most promising are in the fields of microelectronics, photonics and biomedical imaging. Since it has been demonstrated with other more complex techniques that birefringence and, in general, SoP modification induced by biological and electronics samples can indicate their status (e.g., the healthy state of some cells [54–56,69,70] or the stress–strain induced in some materials [15,16]), and considering the achievements of PS-DHI in microelectronics, photonics, and biomedical imaging of the past few years, new technological developments, such as the use of quantum holography, which is a recent fascinating line of research, and new potential applications are expected in the next years.

Author Contributions: Conceptualization, M.A.F. and G.C.; investigation, M.A.F. and G.C.; data curation, M.A.F. and G.C.; writing—original draft preparation, M.A.F. and G.C.; writing—review and editing, M.A.F. and G.C.; supervision, M.A.F. All authors have read and agreed to the published version of the manuscript.

Funding: This research received no external funding.

Conflicts of Interest: The authors declare no conflict of interest. The funders had no role in the design of the study; in the collection, analyses, or interpretation of data; in the writing of the manuscript, or in the decision to publish the results.

Appendix A

Polarization of Light

Polarization of light defines the geometrical orientation of the oscillations of electromagnetic waves. In a transverse wave, the oscillation is in the perpendicular direction respect to the direction of propagation of the beam. When the field vector components along the x and y directions generates a linear trajectory over the time, polarization is called linear, whereas when the tip of the field vector describes a circle or an ellipse in any fixed plane intersecting, and normal to, the direction of propagation, the polarization is classified as circular or elliptical, respectively. The rotation can occur in two possible directions: right circular polarization if the fields rotate in a right-hand sense with respect to the propagation direction, or left circular polarization if the fields rotate in a left-hand sense. Polarization of light can be described with the following two different formalisms [77]:

- Jones vector, evaluated by means of the Jones calculus, only applicable to light that is already fully polarized;
- Stokes parameters, evaluated by means of the Müeller calculus, for a light that is randomly polarized, partially polarized, or incoherent.

Basically, in the Jones formalism, considering propagation along the z axis, the electric field can be written as $E = E_x + E_y$, where

$$\begin{pmatrix} E_x(z,t) \\ E_y(z,t) \end{pmatrix} = \begin{pmatrix} A_x \cos(\omega t - kz + \delta_x) \\ A_y \cos(\omega t - kz + \delta_y) \end{pmatrix} = \begin{pmatrix} A_x e^{i\delta_x} \\ A_y e^{i\delta_y} \end{pmatrix} e^{i(\omega t - kz)} \quad (A1)$$

Here i is the imaginary unit. The components can be writing as a column vector, which is called Jones vector.

$$J = \begin{pmatrix} A_x e^{i\delta_x} \\ A_y e^{i\delta_y} \end{pmatrix} \quad (A2)$$

The state of polarization of an optical wave can be expressed in terms of the amplitudes (A_x, A_y) and the phase variations (δ_x, δ_y) of the x and y components of the electric field vector. Hence, the polarization state of a light beam is completely described by the complex amplitudes in Equation (A2). When a polarized wave with field vector E is incident on a polarization-changing object, the emerging wave has another polarization state E_1 given by

$$\begin{pmatrix} E_{1x} \\ E_{1y} \end{pmatrix} = \begin{pmatrix} j_{11} & j_{12} \\ j_{21} & j_{22} \end{pmatrix} \begin{pmatrix} E_x \\ E_y \end{pmatrix} \quad (A3)$$

where the 2 × 2 transformation matrix is called the Jones matrix. If the optical wave travels through different optical components, the resulting Jones vector can be evaluated by multiplying a cascade of

Jones matrices to the input vector, $J_N J_{N-1} \ldots J_2 J_1 E$, where J_i represents the polarization properties of i-th element [78].

In the case of Stokes formalism, the polarization of light is described by four factors related to intensity and polarization ellipse parameters as described in Figure A1 and in the following equations:

$$\begin{aligned} S_0 &= I \\ S_1 &= Ip \cos 2\psi \cos 2\chi \\ S_2 &= Ip \cos 2\chi \\ S_3 &= Ip \sin 2\chi \end{aligned} \quad (A4)$$

where Ip, 2ψ and 2χ are the spherical coordinates of the polarization state in the three-dimensional space for the S_1, S_2 and S_3 parameters, I is the total intensity of the beam, and p is the degree of polarization given by $\frac{\sqrt{S_1^2 + S_2^2 + S_3^2}}{S_0}$, constrained by $0 \leq p \leq 1$. Generally, normalized Stokes vector, obtained by normalizing to the total intensity S_0, is used and the three significant Stokes parameters are plotted on a spherical region. The parameter S_1 describes the dominance of linear horizontal polarized (LHP) light over linear vertical polarized (LVP) light; S_2 describes the preponderance of linear +45° polarized (L + 45P) light over linear −45° polarized (L − 45P) light and S_3 describes the dominance of right circular polarized (RCP) light over left circular polarized (LCP) light [79]. For pure polarization states, the normalized vector is situated on the Poincaré sphere with unity-radius, while in case of partially polarized states the normalized vector will be placed inside the unity radius Poincaré sphere at a distance of p from the origin.

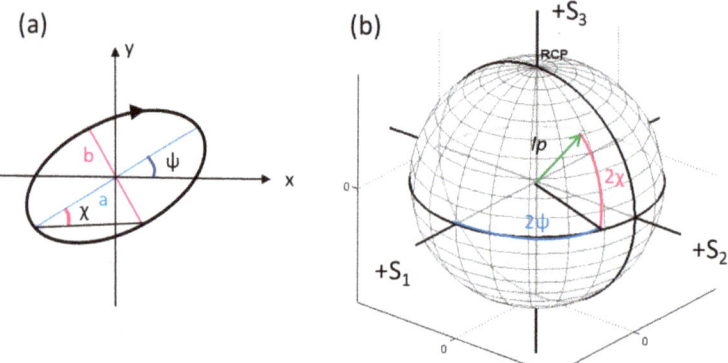

Figure A1. (a) Polarization ellipse, showing the relationship to the Poincaré sphere parameters ψ and χ. In particular, the orientation angle ψ is the angle between the major axis of the ellipse and the x-axis along with the ellipticity $\varepsilon = a/b$, the ratio of the ellipse's major to the minor axis (also known as the axial ratio). The ellipticity parameter is an alternative parameterization of the ellipticity angle, $\chi = \arctan(b/a) = \arctan(1/\varepsilon)$ [15,79]. (b) The Poincaré sphere is the parameterization of the last three Stokes' parameters in spherical coordinates.

The orientation and ellipticity angles, ψ and χ, associated with the polarization ellipse can be related to the Stokes parameters associated with the Poincaré sphere as follows [80]:

$$\begin{aligned} \psi &= \tfrac{1}{2} \tan^{-1}\left(\tfrac{S_2}{S_1}\right), & 0 \leq \psi \leq \pi \\ \chi &= \tfrac{1}{2} \sin^{-1}\left(\tfrac{S_3}{S_0}\right), & -\tfrac{\pi}{4} \leq \chi \leq \tfrac{\pi}{4} \end{aligned} \quad (A5)$$

Following the Müeller formalism, the Stokes components at the output of a given optical sample can be modelled as $S' = M_{sample} S$, where S and S' are the input and output Stokes vectors, respectively, and M is the 4×4 Müeller matrix of the sample,

$$\begin{pmatrix} S'_0 \\ S'_1 \\ S'_2 \\ S'_3 \end{pmatrix} = \begin{pmatrix} M_{11} & M_{12} & M_{13} & M_{14} \\ M_{21} & M_{22} & M_{23} & M_{24} \\ M_{31} & M_{32} & M_{33} & M_{34} \\ M_{41} & M_{42} & M_{43} & M_{44} \end{pmatrix} \begin{pmatrix} S_0 \\ S_1 \\ S_2 \\ S_3 \end{pmatrix} \quad (A6)$$

Similarly to the Jones calculus, when a polarized light wave passes through several optical objects, the polarization of the outgoing beam can be evaluated by knowing the Stokes vector of the input wave and applying Müeller calculus that needs a Müller matrix for each crossed optical object; the resulting vector contains the Stokes parameters of the light outgoing the system [80].

References

1. Buraga-Lefebre, C.; Coetmellec, S.; Lebrun, D.; Ozkul, C. Application of wavelet transform to hologram analysis: Three–dimensional location of particles. *Opt. Laser Eng.* **2000**, *33*, 409–421. [CrossRef]
2. Seebacher, S.; Osten, W.; Baumbach, T.; Jüptner, W. The determination of material parameters of microcomponents using digital holography. *Opt. Laser Eng.* **2001**, *36*, 103–126. [CrossRef]
3. Xu, W.; Jericho, M.H.; Meinertzhagen, I.A.; Kreuser, H.J. Digital in-line holograph microspheres. *Appl. Opt.* **2002**, *41*, 5367–5375. [CrossRef]
4. Ferraro, P.; Grilli, S.; Alfieri, D.; De Nicola, S.; Finizio, A.; Pierattini, A.; Javidi, B.; Coppola, G.; Striano, V. Extended focused image in microscopy by digital holography. *Opt. Express* **2005**, *13*, 6738–6749. [CrossRef]
5. Ferraro, P.; Grilli, S.; Coppola, G.; Javidi, B.; De Nicola, S. How to Extend Depth of Focus in 3D Digital Holography. *Proc. SPIE Three-Dimensional TV Video Display IV* **2005**, *6016*, 60160I. [CrossRef]
6. Nazarathy, M.; Shamir, J. Fourier optics described by operator algebra. *J. Opt. Soc. Amer. A* **1980**, *70*, 150–151. [CrossRef]
7. Cuche, E.; Marquet, P.; Dahlgren, P.; Depeursinge, C.; Delacrétaz, G.; Salathé, R.P. Simultaneous amplitude and quantitative phase-contrast microscopy by numerical reconstruction of Fresnel off-axis holograms. *Appl. Opt.* **1999**, *38*, 6994–7001. [CrossRef]
8. Di Caprio, G.; El Mallahi, A.; Ferraro, P.; Dale, R.; Coppola, G.; Dale, B.; Coppola, G.; Dubois, F. 4D tracking of clinical seminal samples for quantitative characterization of motility parameters. *Biomed. Opt. Express* **2014**, *5*, 690–700. [CrossRef]
9. Coppola, G.; Striano, V.; Ferraro, P.; De Nicola, S.; Finizio, A.; Pierattini, G.; Maccagnani, P. A non-destructive dynamic characterization of a micro-heater through a Digital Holography Microscopy. *J. Microelectromech. Syst.* **2007**, *16*, 659–667. [CrossRef]
10. Dardikman-Yoffe, G.; Mirsky, S.K.; Barnea, I.; Shaked, N.T. High-resolution 4-D acquisition of freely swimming human sperm cells without staining. *Sci. Adv.* **2020**, *6*, eaay7619. [CrossRef]
11. McReynolds, N.; Cooke, F.G.M.; Chen, M.; Powis, S.J.; Dholakia, K. Multimodal discrimination of immune cells using a combination of Raman spectroscopy and digital holographic microscopy. *Sci. Rep.* **2017**, *7*, 43631. [CrossRef]
12. Ferrara, M.A.; De Angelis, A.; De Luca, A.C.; Coppola, G.; Dale, B.; Coppola, G. Simultaneous Holographic Microscopy and Raman Spectroscopy Monitoring of Human Spermatozoa Photodegradation. *IEEE J. Sel. Top. Quantum Electron.* **2016**, *22*, 5200108. [CrossRef]
13. Ferrara, M.A.; De Tommasi, E.; Coppola, G.; De Stefano, L.; Rea, I.; Dardano, P. A New Imaging Method Based on Combined Microscopies. *Int. J. Mol. Sci.* **2016**, *17*, 1645. [CrossRef]
14. Tuchin, V.V. Polarized light interaction with tissues. *J. Biomed. Opt.* **2016**, *21*, 071114. [CrossRef]
15. Colomb, T.; Dahlgren, P.; Beghuin, D.; Cuche, E.; Marquet, P.; Depeursinge, C. Polarization imaging by use of digital holography. *Appl. Opt.* **2002**, *41*, 27–37. [CrossRef]
16. Colomb, T.; Dürr, F.; Cuche, E.; Marquet, P.; Limberger, H.G.; Salathé, R.-P.; Depeursinge, C. Polarization microscopy by use of digital holography: Application to optical-fiber birefringence measurements. *Appl. Opt.* **2005**, *44*, 4461–4469. [CrossRef]

17. Beghuin, D.; Cuche, E.; Dahlgren, P.; Depeursinge, C.; Delacretaz, G.; Salathé, R.P. Single acquisition polarization imaging with digital holography. *Electron. Lett.* **1999**, *35*, 2053–2055. [CrossRef]
18. Palacios, F.; Font, O.; Palacios, G.; Ricardo, J.; Escobedo, M.; Ferreira Gomes, L.; Vasconcelos, I.; Muramatsu, M.; Soga, D.; Prado, A.; et al. Phase and Polarization Contrast Methods by Use of Digital Holographic Microscopy: Applications to Different Types of Biological Samples. In *Holography–Basic Principles and Contemporary Applications*; Mihaylova, E., Ed.; InTech: London, UK, 2013; pp. 353–377.
19. Kemao, Q.; Hong, M.; Xiaoping, W. Real-time polarization phase shifting technique for dynamic deformation measurement. *Opt. Lasers Eng.* **1999**, *31*, 289–295. [CrossRef]
20. Umeda, N.; Iijima, H.; Ishikawa, M.; Takayanagi, A. Birefringence imaging with illumination mode near field scanning optical microscope. In *Far- and Near-Field Optics: Physics and Information Processing*; Jutamulia, S., Asakura, T., Eds.; SPIE digital library: Bellingham, WA, USA, 1998; Volume 3467, pp. 13–17.
21. De Boer, J.F.; Milner, T.E.; van Gemert, M.J.C.; Nelson, J.S. Two-dimensional birefringence imaging in biological tissue by polarization-sensitive optical coherence tomography. *Opt. Lett.* **1997**, *22*, 934–936. [CrossRef] [PubMed]
22. Everett, M.J.; Schoenenberger, K.; Colston, B.W., Jr.; Da Silva, L.B. Birefringence characterization of biological tissue by use of optical coherence tomography. *Opt. Lett.* **1998**, *23*, 228–230. [CrossRef] [PubMed]
23. Oldenbourg, R.; Mei, G. New polarized light microscope with precision universal compensator. *J. Microsc.* **1995**, *180*, 140–147. [CrossRef] [PubMed]
24. Bickel, W.S.; Bailey, W.M. Stokes vectors, Mueller matrices, and polarized scattered light. *Am. J. Phys.* **1985**, *53*, 468–478. [CrossRef]
25. Laskina, A.; Laskina, V.; Ostrun, A. Beam shaping for holographic techniques. *Proc. Spie Opt. Eng. Appl.* **2014**, *9200*, 92000E. [CrossRef]
26. Ohtsuka, Y.; Oka, K. Contour mapping of the spatiotemporal state of polarization of light. *Appl. Opt.* **1994**, *33*, 2633–2636. [CrossRef]
27. Yang, T.D.; Park, K.; Kang, Y.G.; Lee, K.J.; Kim, B.-M.; Choi, Y. Single-shot digital holographic microscopy for quantifying a spatially-resolved Jones matrix of biological specimens. *Opt. Express* **2016**, *24*, 29302–29311. [CrossRef]
28. Liu, X.; Yang, Y.; Han, L.; Guo, C.S. Fiber-based lensless polarization holography for measuring Jones matrix parameters of polarization-sensitive materials. *Opt. Express* **2017**, *25*, 7288–7299. [CrossRef]
29. Kosmeier, S.; Langehanenberg, P.; von Bally, G.; Kemper, B. Reduction of parasitic interferences in digital holographic microscopy by numerically decreased coherence length. *Appl. Phys. B* **2012**, *106*, 107–115. [CrossRef]
30. De Angelis, A.; Ferrara, M.A.; Coppola, G.; Di Matteo, L.; Siani, L.; Dale, B.; Coppola, G.; De Luca, A.C. Combined Raman and polarization sensitive holographic imaging for a multimodal label-free assessment of human sperm function. *Sci. Rep.* **2019**, *9*, 4823. [CrossRef]
31. Colomb, T.; Cuche, E.; Montfort, F.; Marquet, P.; Depeursinge, C. Jones vector imaging by use of digital holography: Simulation and experimentation. *Opt. Commun.* **2004**, *231*, 137–147. [CrossRef]
32. Shimomura, A.; Fukuda, T.; Emoto, A. Analysis of interference fringes based on three circularly polarized beams targeted for birefringence distribution measurements. *Appl. Opt.* **2018**, *57*, 7318–7324. [CrossRef]
33. Berger, V.; Gauthier-Lafaye, O.; Costard, E. Photonic band gaps and holography. *J. Appl. Phys.* **1997**, *82*, 61–64. [CrossRef]
34. Berry, H.G.; Gabrielse, G.; Livingston, A.E. Measurement of the Stokes parameters of light. *Appl. Opt.* **1977**, *16*, 3200–3205. [CrossRef]
35. Azzam, R.M.A. Rotating-detector ellipsometer for measurement of the state of polarization of light. *Opt. Lett.* **1985**, *10*, 427–429. [CrossRef] [PubMed]
36. Jellison, G.E. Four-channel polarimeter for time-resolved ellipsometry. *Opt. Lett.* **1987**, *12*, 766–768. [CrossRef] [PubMed]
37. Williams, P.A. Rotating-wave-plate Stokes polarimeter for differential group delay measurements of polarization-mode dispersion. *Appl. Opt.* **1999**, *38*, 6508–6515. [CrossRef] [PubMed]
38. Oka, K.; Kaneko, T. Compact complete imaging polarimeter using birefringent wedge prisms. *Opt. Express* **2003**, *11*, 1510–1519. [CrossRef]

39. Liu, S.; Han, L.; Li, P.; Zhang, Y.; Cheng, H.; Zhao, J. A method for simultaneously measuring polarization and phase of arbitrarily polarized beams based on Pancharatnam-Berry phase. *Appl. Phys. Lett.* **2017**, *110*, 171112. [CrossRef]
40. Qi, S.; Liu, S.; Li, P.; Han, L.; Zhong, J.; Wei, B.; Cheng, H.; Guo, X.; Zhao, J. A method for fast and robustly measuring the state of polarization of arbitrary light beams based on Pancharatnam-Berry phase. *J. Appl. Phys.* **2019**, *126*, 133105. [CrossRef]
41. Pancharatnam, S. Generalized theory of interference and its applications. *Proc. Ind. Acad. Sci. A* **1956**, *44*, 247–262. [CrossRef]
42. Berry, M.V. The Adiabatic Phase and Pancharatnam's Phase for Polarized Light. *J. Mod. Opt.* **1987**, *34*, 1401–1407. [CrossRef]
43. Milione, G.; Sztul, H.I.; Nolan, D.A.; Alfano, R.R. Higher-order Poincaré sphere, stokes parameters, and the angular momentum of light. *Phys. Rev. Lett.* **2011**, *107*, 053601. [CrossRef] [PubMed]
44. Malhotra, T.; Gutiérrez-Cuevas, R.; Hassett, J.; Dennis, M.R.; Vamivakas, A.N.; Alonso, M.A. Measuring Geometric Phase without Interferometry. *Phys. Rev. Lett.* **2018**, *120*, 233602. [CrossRef] [PubMed]
45. Hariharan, P. *Optical Holography: Principles, Techniques and Applications*, 2nd ed.; Cambridge University Press: New York, NY, USA, 1996; p. 406.
46. Cuche, E.; Bevilacqua, R.; Depeursinge, C. Digital holography for quantitative phase contrast imaging. *Opt. Lett.* **1999**, *24*, 291–293. [CrossRef] [PubMed]
47. Dou, J.; Xi, T.; Ma Jianglei Di, C.; Zhao, J. Measurement of full polarization states with hybrid holography based on geometric phase. *Opt. Express* **2019**, *27*, 7968–7978. [CrossRef] [PubMed]
48. De Zela, F. The Pancharatnam-Berry Phase: Theoretical and Experimental Aspects. In *Theoretical Concepts of Quantum Mechanics*; Reza Pahlavani, M., Ed.; InTech: Rijeka, Croatia, 2012; pp. 289–312.
49. Park, J.; Yu, H.; Park, J.-H.; Park, Y.K. LCD panel characterization by measuring full Jones matrix of individual pixels using polarization-sensitive digital holographic microscopy. *Opt. Express* **2014**, *22*, 24304–24311. [CrossRef]
50. Dubois, F.; Requena, M.-L.N.; Minetti, C.; Monnom, O.; Istasse, E. Partial spatial coherence effects in digital holographic microscopy with a laser source. *Appl. Opt.* **2004**, *43*, 1131–1139. [CrossRef]
51. Ahmad, A.; Dubey, V.; Singh, G.; Singh, V.; Mehta, D.S. Quantitative phase imaging of biological cells using spatially low and temporally high coherent light source. *Opt. Lett.* **2016**, *41*, 1554–1557. [CrossRef]
52. Zhu, Z.; Shi, Z. Self-interference polarization holographic imaging of a three-dimensional incoherent scene. *Appl. Phys. Lett.* **2016**, *109*, 091104. [CrossRef]
53. Röhrich, R.; Hoekmeijer, C.; Osorio, C.I.; Koenderink, A.F. Quantifying single plasmonic nanostructure far-fields with interferometric and polarimetric k-space microscopy. *Light Sci. Appl.* **2018**, *7*, 65. [CrossRef]
54. Wolman, M. Polarized light microscopy as a tool of diagnostic pathology, a review. *J. Histochem. Cytochem.* **1975**, *23*, 21–50. [CrossRef]
55. Chin, L.; Yang, X.; McLaughlin, R.A.; Noble, P.; Sampson, D. Birefringence imaging for optical sensing of tissue damage. In Proceedings of the IEEE Eighth International Conference on Intelligent Sensors, Sensor Networks and Information Processing, Melbourne, Australia, 2–5 April 2013; Volume 1, pp. 45–48. [CrossRef]
56. Chen, H.W.; Huang, C.L.; Lo, Y.L.; Chang, Y.R. Analysis of optically anisotropic properties of biological tissues under stretching based on differential Mueller matrix formalism. *J. Biomed. Opt.* **2017**, *22*, 35006. [CrossRef] [PubMed]
57. Van Turnhout, M.C.; Kranenbarg, S.; van Leeuwen, J.L. Modeling optical behavior of birefringent biological tissues for evaluation of quantitative polarized light microscopy. *J. Biomed. Opt.* **2009**, *14*, 054018. [CrossRef] [PubMed]
58. Shin, I.H.; Shin, S.-M.; Kim, D.Y. New, simple theory-based, accurate polarization microscope for birefringence imaging of biological cells. *J. Biomed. Opt.* **2010**, *15*, 016028. [CrossRef] [PubMed]
59. Low, J.C.M.; Ober, T.J.; Mckinley, G.H.; Stankovic, K.M. Quantitative polarized light microscopy of human cochlear sections. *Biomed. Opt. Exp.* **2015**, *6*, 599–606. [CrossRef] [PubMed]
60. Swami, M.K.; Manhas, S.; Buddhiwant, P.; Ghosh, N.; Uppal, A.; Gupta, P.K. Polar decomposition of 3×3 Mueller matrix: A tool for quantitative tissue polarimetry. *Opt. Express* **2006**, *14*, 9324–9337. [CrossRef]
61. Ghosh, N.; Vitkin, I.A. Tissue polarimetry: Concepts, challenges, applications, and outlook. *J. Biomed. Opt.* **2011**, *16*, 110801. [CrossRef]

62. Phan, Q.-H.; Lo, Y.-L. Stokes-Mueller matrix polarimetry system for glucose sensing. *Opt. Lasers Eng.* **2017**, *92*, 120–128. [CrossRef]
63. Liu, W.-C.; Lo, Y.-L.; Phan, Q.-H. Circular birefringence/dichroism measurement of optical scattering samples using amplitude-modulation polarimetry. *Opt. Lasers Eng.* **2018**, *102*, 45–51. [CrossRef]
64. Chipman, R.A.; Lu, S.Y. Interpretation of Mueller matrices based on polar decomposition. *J. Opt. Soc. Am. A* **1996**, *13*, 1106–1113.
65. Jiao, S.; Yao, G.; Wang, L.V. Depth-resolved two-dimensional Stokes vectors of backscattered light and Mueller matrices of biological tissue measured with optical coherence tomography. *Appl. Opt.* **2000**, *39*, 6318–6324. [CrossRef]
66. He, H.; He, C.; Chang, J.; Lv, D.; Wu, J.; Duan, C.; Zhou, Q.; Zeng, N.; He, Y.; Ma, H. Monitoring microstructural variations of fresh skeletal muscle tissues by Mueller matrix imaging. *J. Biophotonics* **2017**, *10*, 664–673. [CrossRef]
67. Wang, J.; Dong, L.; Chen, H.; Huang, S. Birefringence measurement of biological tissue based on polarizationsensitive digital holographic microscopy. *Appl. Phys. B* **2018**, *124*, 240. [CrossRef]
68. Coppola, G.; Zito, G.; De Luca, A.C.; Ferrara, M.A. Polarized Digital Holography as Valuable Analytical Tool in Biological and Medical Research. In *Digital Holography and Three-Dimensional Imaging*; OSA Technical Digest (Optical Society of America, 2019): Washington, DC, USA, 2019; paper Th4A.5. [CrossRef]
69. Magli, M.C.; Crippa, A.; Muzii, L.; Boudjema, E.; Capoti, A.; Scaravelli, G.; Ferraretti, A.P.; Gianaroli, L. Head birefringence properties are associated with acrosome reaction, sperm motility and morphology. *Reprod. Biomed. Online* **2012**, *24*, 352–359. [CrossRef]
70. Gianaroli, L.; Magli, M.C.; Ferraretti, A.P.; Crippa, A.; Lappi, M.; Capitani, S.; Baccetti, B. Birefringence characteristics in sperm heads allow for the selection of reacted spermatozoa for intracytoplasmic sperm injection. *Fertil. Steril.* **2010**, *93*, 807–813. [CrossRef] [PubMed]
71. Coppola, G.; Di Caprio, G.; Wilding, M.; Ferraro, P.; Esposito, G.; Di Matteo, L.; Dale, R.; Coppola, G.; Dale, B. Digital holographic microscopy for the evaluation of human sperm structure. *Zygote* **2014**, *22*, 446–454. [CrossRef] [PubMed]
72. Edengeiser, E.; Meister, K.; Bründermann, E.; Büning, S.; Ebbinghaus, S.; Havenith, M. Non-invasive chemical assessment of living human spermatozoa. *RSC Adv.* **2005**, *5*, 10424–10429. [CrossRef]
73. Gordon, G.S.; Joseph, J.; Sawyer, T.; Macfaden, A.J.; Williams, C. Full-field quantitative phase and polarisation-resolved imaging through an optical fibre bundle. *Opt. Express* **2019**, *27*, 23929–23947. [CrossRef] [PubMed]
74. Öhman, J.; Gren, P.; Sjödahl, M. Polarization-resolved dual-view holographic system for 3D inspection of scattering particles. *Appl. Opt.* **2019**, *58*, G31–G40. [CrossRef]
75. Öhman, J.; Sjödahl, M. Identification, tracking, and sizing of nano-sized particles using dual-view polarization-resolved digital holography and T-matrix modeling. *Appl. Opt.* **2020**, *59*, 4548–4556.
76. Bouchal, P.; Štrbková, L.; Dostál, Z.; Chmelík, R.; Bouchal, Z. Geometric-Phase microscopy for quantitative phase imaging of isotropic, birefringent and space-variant polarization samples. *Sci. Rep.* **2019**, *9*, 3608. [CrossRef]
77. Perez, J.J.G.; Ossikovski, R. *Polarized Light and the Mueller Matrix Approach*, 1st ed.; CRC Press: Boca Raton, FL, USA, 2016; p. 405.
78. Baumann, B. Polarization Sensitive Optical Coherence Tomography: A Review of Technology and Applications. *Appl. Sci.* **2017**, *7*, 474. [CrossRef]
79. Collett, E. *Field Guide to Polarization*; SPIE Press: Bellingham, WA, USA, 2005; p. 148.
80. Singh, D.K. Propagation of Light: A Review. *Int. J. Res. Sci. Innov.* **2017**, *IV*, 70–74.

© 2020 by the authors. Licensee MDPI, Basel, Switzerland. This article is an open access article distributed under the terms and conditions of the Creative Commons Attribution (CC BY) license (http://creativecommons.org/licenses/by/4.0/).

Article

Analysis of Pulses Bandwidth and Spectral Resolution in Femtosecond Stimulated Raman Scattering Microscopy

Luigi Sirleto [1], Rajeev Ranjan [1,2] and Maria Antonietta Ferrara [1,*]

[1] National Research Council (CNR), Institute of Applied Sciences and Intelligent Systems, Via Pietro Castellino 111, 80131 Naples, Italy; luigi.sirleto@cnr.it (L.S.); rajeev.ranjan@iit.it (R.R.)
[2] CHT @Erzelli Nanoscopy Istituto Italiano di Tecnologia, 16152 Genova, Italy
* Correspondence: antonella.ferrara@na.isasi.cnr.it

Featured Application: Stimulated Raman microscopy, based on two femtosecond pulsed lasers and with a spectral resolution of about 56 cm^{-1}, is demonstrated to be sufficient in order to distinguish protein and lipid bands in the C-H region.

Abstract: In the last decade, stimulated Raman scattering (SRS) imaging has been demonstrated to be a powerful method for label-free, non-invasive mapping of individual species distributions in a multicomponent system. This is due to the chemical selectivity of SRS techniques and the linear dependence of SRS signals on the individual species concentrations. However, even if significant efforts have been made to improve spectroscopic coherent Raman imaging technology, what is the best way to resolve overlapped Raman bands in biological samples is still an open question. In this framework, spectral resolution, i.e., the ability to distinguish closely lying resonances, is the crucial point. Therefore, in this paper, the interplay among pump and Stokes bandwidths, the degree of chirp-matching and the spectral resolution of femtosecond stimulated Raman scattering microscopy are experimentally investigated and the separation of protein and lipid bands in the C-H region, which are of great interest in biochemical studies, is, in principle, demonstrated.

Keywords: stimulated Raman microscopy; pulsed source; laser pulse bandwidths; laser chirping; spectral resolution

1. Introduction

Over the past ten years, stimulated Raman scattering (SRS) microscopy has been investigated in nanophotonics [1–4] as well as in biophotonics as an analytical, label-free, non-invasive technique with unique cellular and tissue imaging capabilities [5–8]. As almost all the biomolecules contain carbon and hydrogen, the CH-stretching (2800–3100 cm^{-1}) region of Raman spectra of biomolecules is the most used in SRS microscopy. The two Raman bands typically investigated are CH_2, near to 2845, and CH_3, near to 2930 cm^{-1}, corresponding to lipids and proteins, respectively. Due to their large spectral shapes (about 100 cm^{-1}) and the difference between peaks of 95 cm^{-1}, the CH_2 and CH_3 are partially overlapped. Moreover, in the C-H region, the SRS signal level is high because the density of CH bonds is high, while the SRS molecular specificity is assumed to be low.

Typically, SRS microscopy is implemented by using two Fourier transform-limited (FTL) tunable picosecond (ps) laser sources with a high spectral resolution ($\approx$10 cm^{-1}), helpful in the region of interest (i.e., the fingerprint region: 800 ÷ 1800 cm^{-1}), where Raman peaks are narrow, nearly spaced, and could be packed [8]. However, when ps laser pulses are used to implement an SRS microscope, an equally fruitful imaging in carbon–hydrogen (C-H) stretching is achieved. The drawback is that ps pulses show a low peak intensity, thus needing high laser power for imaging [5–8]. In the last decade, an improvement of about one order of magnitude of the SRS signal has been demonstrated when ps pulses are replaced with femtosecond (fs) pulses [9]. This improvement is due to a higher peak intensity; thus, higher signal-to-noise ratio (SNR) can be obtained when temporally shorter pulses are used than narrowband, picosecond pulses with an unchanged optical power. However, when ultra-fast sources are used, a low spectral selectivity is obtained and multi-band excitation can occur, not allowing, in principle, the separation of some bands of particular interest in biology, such as those of lipids and proteins as discussed previously. To solve this issue, a number of methods for SRS multicolor imaging, based on broadband femtosecond pulses, have been developed. Among them, frequency tuning [10], multiplexing [11,12] and spectral focusing [13–16] implementations have been studied and reported in the literature.

It is well known that the light pulse is considered transform-limited when the angular frequency is constant and equals the central angular frequency $w(t) = w_0$. On the other hand, the chirp of an optical pulse is generally understood as the time dependence of its instantaneous frequency; thus, a chirped pulse having a carrier frequency w_0 at time t shows an instantaneous central frequency $w(t)$ that depends on the linear chirp parameter β by the equation: $w(t) = w_0 + 2\beta t$ [17]. In detail, a down-chirp (up-chirp) means that the instantaneous frequency decreases (increases) with time. A pulse can gain a chirp, for example, through propagation in a transparent medium due to the effects of chromatic dispersion and nonlinearities. Indeed, due to its wide spectral width and to group velocity dispersion, optical pulse propagating in a transparent medium undergoes a phase distortion inducing an increase in its duration with a different laser frequency distribution in time, as reported in Figure 1. In particular, the time–bandwidth product $\Delta w_0 \cdot \tau_0 = 4 \ln 2 \approx 2.77$ corresponds to the area of the ellipse on the left in Figure 1 and it is the bandwidth of an FTL laser pulse. As a result of the chirp, laser pulses can undergo a temporal stretch and the final pulse duration is $\tau = F\tau_0$, where τ_0 is the FTL pulse duration and F is the stretching factor; at the same time, the instantaneous spectral bandwidth becomes narrower than the FTL spectral bandwidth by a factor of $1/F$ [18]. With the relation $\Delta w \cdot \tau = 2.77$ also being applicable to the chirped pulse width τ and to the instantaneous bandwidth Δw of the pulse, the duration broadening leads to a decrease in the instantaneous bandwidth by the stretching factor, whereas the whole bandwidth Δw_0 is left unchanged [19], as depicted in Figure 1.

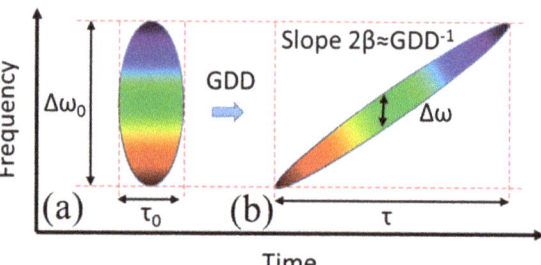

Figure 1. Time-bandwidth distribution for (**a**) Fourier-transform limited laser pulse and (**b**) the same laser pulse that is linearly chirped.

In order to enhance spectral resolution in SRS microscopy based on fs laser pulses, an option is to force a quadratic spectral phase variation. By equally chirping pump and Stokes beams with an energy spacing corresponding to the Raman line, it is possible to generate a constant instantaneous frequency difference (IFD, $\Omega = w_p - w_s$) that spectrally focuses the excitation energy into a single resonance. We note that the bandwidth $\delta\Omega$, which ultimately determines the SRS spectral resolution, in the limiting case can simulate the ps SRS system. This method is known as spectral focusing (SF) [13–16]. Nevertheless, the great disadvantage related to this approach is the large amount of parameters that should be taken into account in the selection and alignment of the optics to obtain the chirp-matching condition. Moreover, since the operative conditions can be altered by fluctuations in the pump and Stokes wavelengths together with the dispersion in the microscope, perfect chirp-matching is difficult to maintain. For these reasons, the resulting spectral resolution of SF-SRS setups is often worse than theoretically predicted [13–16].

The comparison among FTL, equally and differently chirped laser sources is shown in Figure 2. In the spectral focusing, when the pump and Stokes pulses are "chirp-matched", the bandwidth of IFD is lower than the total bandwidths of the transform-limited pulses, leading to an improvement of the spectral resolution, as reported in Figure 2b. On the other hand, when the chirp-matching condition is not met, the bandwidth of the IFD signal is broader, and, as a consequence, the spectral resolution is poorer with respect to the chirp matched case (see Figure 2c). Interestingly, spectral resolution for differently chirped lasers is better than FTL laser sources [19].

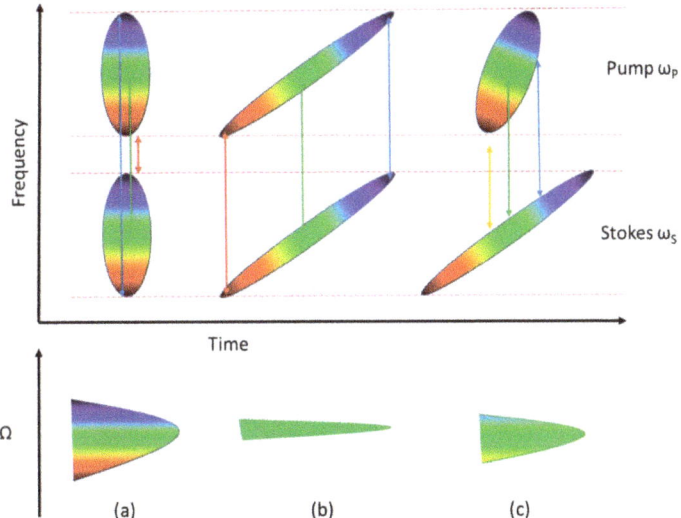

Figure 2. Coherent Raman excitation with (**a**) FTL laser pulses, (**b**) laser pulses with the same chirp, (**c**) laser pulses with different chirps. Ω is the vibration frequency given by $\Omega = \omega_p - \omega_s$.

For microscopy, both contrast imaging and spectral tuning are important to assess the good quality of images. However, in coherent Raman scattering, there is a tradeoff between the best spectral resolution, which is reached with ps sources, and the best ratio of image contrast and signal intensity, which is obtained when the spectral resolution and the width of the Raman lines under observation are almost the same [14]. This condition is satisfied in the case of excitation with both ps pulses, since they match the linewidths in the fingerprint region (5–20 cm^{-1}), and in the case of broader bandwidth femtosecond (fs) pulses, which are considered the ideal excitation for CH stretching vibrations.

We note that the SF approach usually takes advantage by static dispersive elements, such as glass rods, placed into the optical path to stretch both the pump and the Stokes frequencies in time and produce a constant IFD. However, the optical elements embedded in an SRS optical circuitry microscope setup can also change the pulse width. In this paper, taking advantage of the small chirping, introduced by simply propagating the beams through dispersive materials already present in the SRS microscope setup, we examine the spectral correlations and we demonstrate that its value is enough to distinguish protein and lipid bands in the C-H region.

2. Experimental Setup and Methods

Our optical system, not commercially existing and schematically shown in Figure 3, is obtained by integrating a femtosecond-SRS spectroscopy setup [20–22] with a C2 confocal Nikon microscope, which consists of an inverted Nikon Ti-eclipse microscope and a scan head. Experimental details of our setup have already been reported in our previous papers [22–25]. Two femtosecond laser sources are involved in this setup: a Ti:Sapphire (Ti:Sa—Chameleon Ultra II—pulse duration = 140 fs, repetition rate = 80 MHz, emission wavelengths range = 680–1080 nm) and a femtosecond synchronized optical parametric oscillator (OPO—Chameleon Compact OPO—pulse duration = 200 fs, repetition rate = 80 MHz, emission wavelengths range = 1000–1600 nm).

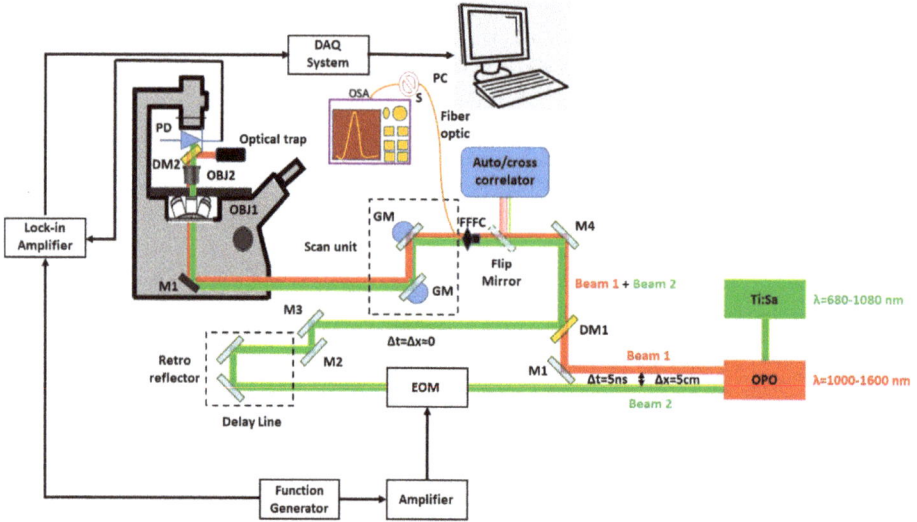

Figure 3. Experimental setup for stimulated Raman scattering imaging. (Ti:Sa): Titanium-Sapphire pulsed laser source; OPO: optical parameter oscillator; M: mirror; EOM: electro-optical modulator; GM: galvo-mirror; DM: dichroic mirror; OBJ: objective lens; PD: photo-detector; FFFC: flip flop fiber coupler; S: multimode optical fiber; OSA: optical spectrum analyzer.

Typically, the temporal characterization of a signal is performed by measuring the correlation between the signal and its duplication. In particular, optical autocorrelation of the field intensity may be used to measure the second-order coherence degree and to assess the duration of short pulses. With this aim, an additional flip-flop mirror has been mounted in-between the mirror M4 and the input of the scan head in our optical architecture to deflect the laser pulse beams directly toward an autocorrelator (pulseCheck 50–A.P.E., Berlin) (see Figure 3); therefore, auto- and cross-correlation of the pulse beam were measured in the optical path just before the laser beam reaches the microscope. In an autocorrelator, the input pulsed beam is divided into two arms of a Michelson interferometer and a delay can be induced in the pulses in one arm with respect to the other one. Both pulses are then overlapped in a non-linear crystal and the generated signal is detected. The pulse duration can be evaluated from the measurement of the time delay and the intensity of the generated signal [26]. The detectable pulse width range is fixed by the delay range, whereas the measurable wavelength range is determined by the detector and the non-linear material. Here, the nonlinear process is the two photons absorption (TPA) and it is measured as a function of the time delay giving the beam autocorrelation function. Auto-correlators based on TPA have some important advantages: (i) a higher sensitivity can be obtained with respect to second harmonic generation (SHG) due to the involvement of fields only at the original frequency, ω, owing to the TPA resonant second-order transition nature; (ii) extremely short pulses can be characterized since TPA can operate in a wide wavelength range not restricted by the narrow phase matching bandwidth; (iii) in TPA, non-linear signal multiplication and detection are combined into one process, leading to a simplification and a higher efficiency with respect to a two-step process of optical non-linearity followed by linear detection.

The used pulseCheck can be used in two measurement modes: collinear and non-collinear. Collinear, also known as interferometric or fringe-resolved mode, provides further qualitative information on the chirp and central wavelength of the pulse. Regarding non-collinear mode, it allows high dynamic range measurements that are background free [27].

The auto- correlator was used with a pulse width measurement range of (10 fs–12 ps) and it was connected to the PC through USB and by using the APE's Standardized Software Interface, allowing either remote control or integration into automated setups. Therefore, the deflected beam is tuned until the intensity is stabilized and maximized, and then, the pulse is acquired and the data analysis is completed by using the Gaussian fitting curve function *cftool* of MATLAB 2020.

To evaluate the spectral resolution, the exact estimation of the pulses' duration is required. Normally, the full width at half-maximum (FWHM) of the unknown pulse τ_p is proportional to the FWHM of the measured fringe-resolved intensity autocorrelation function τ_{ac}:

$$\tau_{ac} = k \cdot \tau_p \qquad (1)$$

where k is the proportionality factor, also known as the deconvolution factor. k differs significantly for different pulse shapes; therefore, to evaluate the pulse width from the intensity autocorrelation requires some previous knowledge of the pulse shape. Typically, the deconvolution factor can be calculated for analytical pulse shapes or computed numerically for complicated pulses; however, for some common pulse shapes, the deconvolution factor is known (k = 1.414 for Gaussian shape, k = 1.543 for sech shape, k = 1 for square shape) [28].

It is well known that the Fourier transformation correlates β, which defines the linear slope of the central frequency, to the group delay dispersion (GDD) applied to the laser pulse as reported in the following [19]:

$$\beta = \frac{2GDD}{\tau_0^4 + 4GDD^2} \approx \frac{1}{2GDD} \; for \; \tau \gg \tau_0 \qquad (2)$$

The GDD leads to a stretch of the pulse width, which passes from τ_0 in the case of the FTL pulse to the chirped width τ:

$$\tau = \tau_0 \sqrt{1 + \left(\frac{4\ln 2 \cdot GDD}{\tau_0^2}\right)^2} \approx 2.77 \frac{|GDD|}{\tau_0} \; for \; \tau \gg \tau_0 \qquad (3)$$

From Equations (2) and (3), we found that $\tau \cdot \tau_0 \approx 4\ln 2 |GDD| \approx 2\ln 2/|\beta|$; this product is considered a direct measure of the chirp parameter β for $\tau \gg \tau_0$ [19]. We note that two chirped pulses have the same slope only if the resultant products $\tau \cdot \tau_0$ are equal.

Information obtained by autocorrelation measurements allows us to monitor the pulse duration and chirp of the laser beam, which are very important parameters to optimize the non-linear interaction in microscopy. Moreover, considering that SRS is a two-pulse technique, its spectral resolution is not defined by the spectrum of the individual exciting pulses but by the spectrum of their temporal interference. Consequently, cross-correlation characterization is equally important for providing information about the entire system; in particular, the FWHM of pump and probe beams' cross-correlation allows us to evaluate the experimental spectral bandwidth [29]. Typically, SF is obtained by equally chirping the pump and the Stokes pulses by using glass elements of known group-velocity dispersion without significant intensity losses [30]. As a general rule, the spectral resolution is restricted both by the level to which the pulses are chirped and by the similarity of these chirps. A better spectral resolution can be obtained when the pump and Stokes pulses with larger bandwidths are chirp-matched. In the case of transform limited laser pulses, the spectral resolution can be calculated by the formula [19]:

$$\Delta \widetilde{\nu} = \frac{2\ln 2}{\pi c}\sqrt{2\left(\tau_p^{-2} + \tau_S^{-2}\right)} = 20.8 \text{ ps·cm}^{-1}\sqrt{\tau_p^{-2} + \tau_S^{-2}} \qquad (4)$$

and in the case of our fs laser sources its evaluated value is of 181 cm^{-1}. In order to carry out cross-correlation between Ti:Sa and OPO, both sources are focused inside the TPA detector, then by inserting an optical delay line (Newport MOD MILS200CC) between the Ti:Sa and the microscope, we introduce an optical delay in the Ti:Sa beam; at each step of the delay line, the signal is acquired.

Finally, to have a complete characterization of the pulsed sources used in our SRS microscope, we have carried out laser spectra measurement by adding a flip flop fiber coupler (FFFC in Figure 3) in the optical setup without disturbing it, and the deflected beam is coupled in an optical spectrum analyzer (OSA—Ando AQ6317C) by a multimode optical fiber S, which is mounted after the flip/flop mirror. In Figure 3, the setup used integrated with the described system for complete characterization of the pulsed laser sources is reported.

3. Results and Discussion

Autocorrelation characterizations of the pulsed laser beams were examined using the aforementioned autocorrelator. The Ti:Sa and OPO laser emission wavelengths were fixed at 811 nm and at 1074 nm, respectively. The collected autocorrelation function of the Ti:Sa and OPO is reported in Figure 4a,b, respectively. As can be seen in these figures, the pulses emitted by the lasers show a Gaussian-like distribution, as expected. Therefore, with the aim to evaluate the pulse width value, both the autocorrelation traces were Gaussian-fitted as displayed in Figure 4a,b, and the corresponding FWHM of the Gaussian curves was calculated to be about 341 fs and 357 fs for Ti:Sa and OPO, respectively. Since the pulse follows a Gaussian line shape and considering Equation (1), the factor of 1.414 was applied to our evaluation, leading to an estimation of the pulses' width duration at the input of the microscope of about τ = 241 fs for Ti:Sa and τ = 253 fs for OPO, respectively. Thus, a small broadening of OPO was observed, while a major one was retrieved for Ti:Sa. These broadenings are related to the dispersion arising during propagation through several optical elements along the beams' path; in particular, the higher widening observed for Ti:Sa can be explained considering that in its optical path a Pockels cell is employed. In Table 1, the features of the two lasers and results of their autocorrelation measurements are summarized.

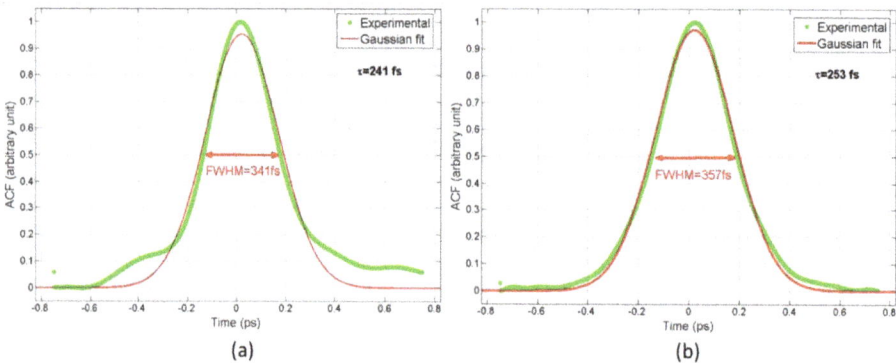

Figure 4. Autocorrelation trace of the pulsed laser sources (**a**) Ti:Sa and (**b**) OPO, respectively. Results of the Gaussian fit are also reported.

Table 1. Pulsed laser sources properties.

Laser Source	Wavelength Range [nm]	Pulse Duration [fs]	Repetition Rate [MHz]	Widened Pulse Duration [fs]
Ti:Sa	740–880	140	80	241
OPO	1000–1600	200	80	253

The second order group velocity dispersion can be evaluated for each pulsed source by applying Equations (2) and (3). The obtained values are GDD = 9905 fs^2 and GDD = 11,177 fs^2 for Ti:Sa and OPO, respectively. In our case, pulses were not equally chirped.

To complete the single beam characterization, the laser spectra were acquired by an optical spectrum analyzer and results are shown in Figure 5, when Ti:Sa is tuned to 811 nm and OPO to 1074 nm, respectively. The resultant bandwidths are approximately the same for both pulsed laser sources and were of about 6.1 nm.

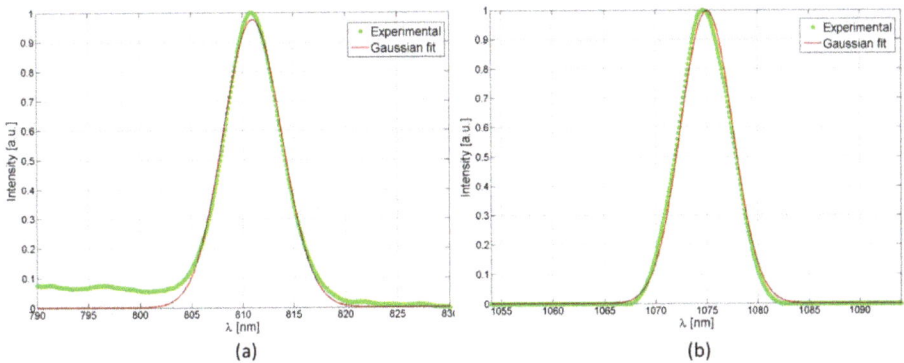

Figure 5. Optical spectra of the pulsed laser sources (**a**) Ti:Sa and (**b**) OPO, respectively. Results of the Gaussian fit are also reported.

The measured FWHM of Ti:Sa and OPO cross-correlation was 371 fs; since the convolution of two Gaussian functions is another a Gaussian function [31], Equation (1) can also be applied when cross-correlation measurements are performed, leading to a pulse duration of 262 fs (see Figure 6). Thus, the obtained experimental spectral bandwidth, which is given by the FWHM of cross-correlation in the frequency domain, was of 56 cm^{-1}.

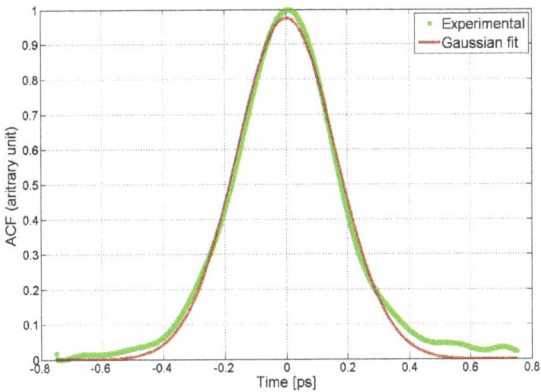

Figure 6. Cross-correlation measure and its Gaussian fit.

The retrieved spectral resolution is very useful in biological sample imaging when both lipid and protein bands need to be investigated by SRS microscopy by simply regulating the frequency either of the pump or the Stokes beams in successive scans. The corresponding two stretching signals are CH_2 (2845 cm^{-1}) and CH_3 (2940 cm^{-1}), respectively; thus, these bands can be collected by choosing one Raman shift at a time, allowing the imaging of the map distributions of the lipid and protein on the same field of the sample. Therefore, since our experimental spectral bandwidth was of 56 cm^{-1}, when we set the laser beams to excite the 2845 cm^{-1} lipid band, we are exciting from (FWHM) 2817 to 2873 cm^{-1}; in the same way, when we set the laser beams to the protein band at 2940 cm^{-1}, effectively we excite the range 2912–2968 cm^{-1} (FWHM). In Figure 7, the spectral bandwidth derived by the chirping of the pulses (56 cm^{-1}—continuous lines) and two simulated Raman bands with two peaks at a distance of 95 cm^{-1} with a width of about 100 cm^{-1} (red circles and green diamonds) are illustrated.

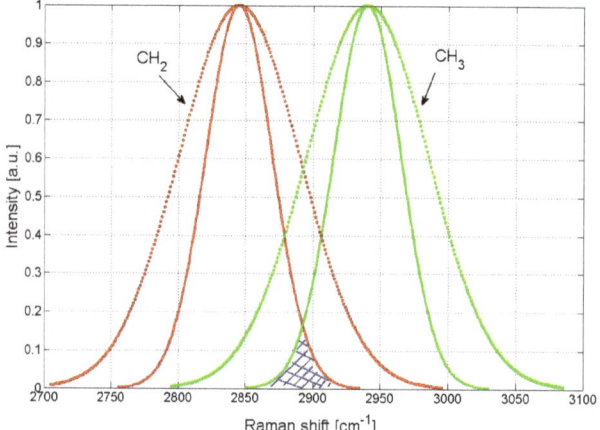

Figure 7. Raman bands for CH_2 (2845 cm^{-1}, red circles) and CH_3 (2940 cm^{-1}, green diamonds). Continuous lines are obtained considering the Ti:Sa and OPO cross-correlation reported in Figure 6 at the input of the microscope (i.e., 262 fs). Blue lines highlight the overlap area between two excited bandwidths.

The overlap between two excited bandwidths is highlighted with blue lines; however, the contribution to the Raman signal from this region can be neglected considering that the intensities are lower than FWHM values and, therefore, they can be considered under the threshold. Thus, we can conclude that, using 56 cm^{-1} chirped pulses, the 2845 cm^{-1} channel is essentially related to the lipid signal and the 2940 cm^{-1} channel is principally due to protein content. Definitely, considering the retrieved experimental spectral resolution (about 56 cm^{-1}) that is lower than the Raman band of lipids and proteins (about 100 cm^{-1}) and since the overlap region, obtained when lipids and proteins are sequentially excited, can be considered negligible, our SRS microscope is suitable to image molecular specificity such as lipids (CH_2) and proteins (CH_3) in the C-H region, as already demonstrated in our previous paper [23].

4. Conclusions

In SRS microscopy, the trade-off between high selectivity offered by picosecond pulses and high SRS signal obtained by using femtosecond pulses is still an open question and it is widely investigated. In particular, for biological application, there is interest in label-free imaging both the lipid and protein distribution. Unfortunately, the Raman bands of these two components are very close, thus ps pulses are required; however, the protein content could be low, leading to very weak Raman signals, so fs pulses would be appropriate.

In this paper, in order to overcome the drawback of spectral focusing, an alternative method is proposed. Its basic idea is to avoid adding optical elements in the SRS microscopy optical setup and to take advantage of chirping, introduced by simply propagating the beams through dispersive materials already present in the SRS microscope. The pros are that no additional optical elements have to be introduced in the experimental setup, giving the great advantage of a more simple and cheap setup; the cons are that the spectral resolution is fixed. However, in our setup, an experimental value of 56 cm^{-1} for spectral resolution is measured by cross-correlation techniques, and molecular specificity is demonstrated for lipids and proteins in the C-H region.

Moreover, the spectral resolution was measured before the microscope, and this means that a significant further chirping of laser pulses is expected due to the propagation of the pulsed beams inside the scan head and of the microscope objective, leading to an additional improvement in spectral resolution. Definitely, our method has the merit to maintain the benefits of femtosecond pulses, i.e., an improvement in sensitivity with respect to ps pulses, while preserving in the C-H region of Raman spectra of biomolecules an adequate spectral resolution.

Author Contributions: Conceptualization, methodology, investigation, validation, data analysis, writing—original draft preparation: M.A.F., R.R. and L.S. All authors have read and agreed to the published version of the manuscript.

Funding: This research received no external funding.

Data Availability Statement: The datasets generated during and/or analyzed during the current study are available from the corresponding author on reasonable request.

Acknowledgments: The authors would like to thank M. Indolfi and V. Tufano (ISASI-CNR) for their precious and constant technical assistance.

Conflicts of Interest: The authors declare no conflict of interest. The funders had no role in the design of the study; in the collection, analyses, or interpretation of data; in the writing of the manuscript, or in the decision to publish the results.

References

1. Sirleto, L.; Ferrara, M.A.; Nikitin, T.; Novikov, S.; Khriachtchev, L. Giant Raman gain in silicon nanocrystals. *Nat. Commun.* **2012**, *3*, 1220. [CrossRef] [PubMed]
2. Sirleto, L.; Vergara, A.; Ferrara, M.A. Advances in stimulated Raman scattering in nanostructures. *Adv. Opt. Photon* **2017**, *9*, 169–217. [CrossRef]
3. Ferrara, M.; Sirleto, L.; Nicotra, G.; Spinella, C.; Rendina, I. Enhanced gain coefficient in Raman amplifier based on silicon nanocomposites. *Photon Nanostruct. Fundam. Appl.* **2011**, *9*, 1–7. [CrossRef]
4. Sirleto, L.; Aronne, A.; Gioffrè, M.; Fanelli, E.; Righini, G.C.; Pernice, P.; Vergara, A. Compositional and thermal treatment effects on Raman gain and bandwidth in nanostructured silica based glasses. *Opt. Mater.* **2013**, *36*, 408–413. [CrossRef]
5. Freudiger, C.W.; Min, W.; Saar, B.G.; Lu, S.; Holtom, G.R.; He, C.; Tsai, J.C.; Kang, J.X.; Xie, X.S. Label-Free Biomedical Imaging with High Sensitivity by Stimulated Raman Scattering Microscopy. *Science* **2008**, *322*, 1857–1861. [CrossRef] [PubMed]
6. Cheng, J.-X.; Xie, X.S. Vibrational spectroscopic imaging of living systems: An emerging platform for biology and medicine. *Science* **2015**, *350*, aaa8870. [CrossRef]
7. Camp, C.H., Jr.; Cicerone, M.T. Chemically sensitive bioimaging with coherent Raman scattering. *Nat. Photon* **2015**, *9*, 295–305. [CrossRef]
8. Zumbusch, A.; Langbein, W.; Borri, P. Nonlinear vibrational microscopy applied to lipid biology. *Prog. Lipid Res.* **2013**, *52*, 615–632. [CrossRef]
9. Zhang, D.; Slipchenko, M.N.; Cheng, J.-X. Highly Sensitive Vibrational Imaging by Femtosecond Pulse Stimulated Raman Loss. *J. Phys. Chem. Lett.* **2011**, *2*, 1248–1253. [CrossRef]
10. Steinle, T.; Kumar, V.; Floess, M.; Steinmann, A.; Marangoni, M.; Koch, C.; Wege, C.; Cerullo, G.; Giessen, H. Synchroniza-tion-free all-solid-state laser system for stimulated Raman scattering microscopy. *Light Sci. Appl.* **2016**, *5*, e16149. [CrossRef] [PubMed]
11. Fu, D.; Lu, F.K.; Zhang, X.; Freudiger, C.W.; Pernik, D.R.; Holtrom, G.; Xie, X.S. Quantitative chemical imaging with mul-tiplex stimulated Raman scattering microscopy. *J. Am. Chem. Soc.* **2012**, *134*, 3623–3626. [CrossRef]
12. Liao, C.-S.; Wang, P.; Wang, P.; Li, J.; Lee, H.J.; Eakins, G.; Cheng, J.-X. Spectrometer-free vibrational imaging by retrieving stimulated Raman signal from highly scattered photons. *Sci. Adv.* **2015**, *1*, e1500738. [CrossRef]

13. Fu, D.; Holtom, G.; Freudiger, C.; Zhang, X.; Xie, X.S. Hyperspectral Imaging with Stimulated Raman Scattering by Chirped Femtosecond Lasers. *J. Phys. Chem. B* **2013**, *117*, 4634–4640. [CrossRef]
14. Brückner, L.; Buckup, T.; Motzkus, M. Exploring the potential of tailored spectral focusing. *J. Opt. Soc. Am. B* **2016**, *33*, 1482–1491. [CrossRef]
15. Pegoraro, A.F.; Ridsdale, A.; Moffatt, D.J.; Jia, Y.W.; Pezacki, J.P.; Stolow, A. Optimally chirped multimodal CARS micros-copy based on a single Ti:sapphire oscillator. *Opt. Express* **2009**, *17*, 2984–2996. [CrossRef]
16. Rocha-Mendoza, I.; Langbein, W.; Borri, P. Coherent anti-Stokes Raman microspectroscopy using spectral focusing with glass dispersion. *Appl. Phys. Lett.* **2008**, *93*, 201103. [CrossRef]
17. Hirlimann, C. Pulsed Optics. In *Femtosecond Laser Pulses Principle and Experiments*, 2nd ed.; Rullier, C., Ed.; Springer: New York, NY, USA, 2005; pp. 25–56.
18. Beier, H.T.; Noojin, G.D.; Rockwell, B.A. Stimulated Raman scattering using a single femtosecond oscillator with flexibility for imaging and spectral applications. *Opt. Express* **2011**, *19*, 18885–18892. [CrossRef] [PubMed]
19. Mohseni, M.; Polzer, C.; Hellerer, T. Resolution of spectral focusing in coherent Raman imaging. *Opt. Express* **2018**, *26*, 10230–10241. [CrossRef] [PubMed]
20. Ranjan, R.; D'Arco, A.; Ferrara, M.A.; Indolfi, M.; Larobina, M.; Sirleto, L. Integration of stimulated Raman gain and stimu-lated Raman losses detection modes in a single nonlinear microscope. *Opt. Express* **2018**, *26*, 26317–26326. [CrossRef]
21. D'Arco, A.; Ferrara, M.A.; Indolfi, M.; Tufano, V.; Sirleto, L. Label-free imaging of small lipid droplets by femtosec-ond-stimulated Raman scattering microscopy. *J. Nonlinear Opt. Phys. Mater.* **2017**, *26*, 1750052. [CrossRef]
22. Ranjan, R.; Indolfi, M.; Ferrara, M.A.; Sirleto, L. Implementation of a Nonlinear Microscope Based on Stimulated Raman Scattering. *J. Vis. Exp.* **2019**, *2019*, e59614. [CrossRef]
23. Ferrara, M.A.; Filograna, A.; Ranjan, R.; Corda, D.; Valente, C.; Sirleto, L. Three-dimensional label-free imaging throughout adipocyte differentiation by stimulated Raman microscopy. *PLoS ONE* **2019**, *14*, e0216811. [CrossRef]
24. Ranjan, R.; Ferrara, M.; Filograna, A.; Valente, C.; Sirleto, L. Femtosecond Stimulated Raman microscopy: Home-built realization and a case study of biological imaging. *J. Instrum.* **2019**, *14*, P09008. [CrossRef]
25. D'Arco, A.; Brancati, N.; Ferrara, M.A.; Indolfi, M.; Frucci, M.; Sirleto, L. Subcellular chemical and morphological analysis by stimulated Raman scattering microscopy and image analysis techniques. *Biomed. Opt. Express* **2016**, *7*, 1853–1864. [CrossRef] [PubMed]
26. Piazza, V.; De Vito, G.; Farrokhtakin, E.; Ciofani, G.; Mattoli, V. Femtosecond-Laser-Pulse Characterization and Optimization for CARS Microscopy. *PLoS ONE* **2016**, *11*, e0156371. [CrossRef]
27. Berlin, A.P.E. Available online: https://www.ape-berlin.de/en/autocorrelator/ (accessed on 10 February 2021).
28. Chen, J.; Xia, W.; Wang, M. Characteristic measurement for femtosecond laser pulses using a GaAs PIN photodiode as a two-photon photovoltaic receiver. *J. Appl. Phys.* **2017**, *121*, 223103. [CrossRef]
29. Ito, T.; Obara, Y.; Misawa, K. Single-beam phase-modulated stimulated Raman scattering microscopy with spectrally focused detection. *J. Opt. Soc. Am. B* **2017**, *34*, 1004. [CrossRef]
30. Andresen, E.R.; Berto, P.; Rigneault, H. Stimulated Raman scattering microscopy by spectral focusing and fiber-generated soliton as Stokes pulse. *Opt. Lett.* **2011**, *36*, 2387–2389. [CrossRef]
31. Getreuer, P. A Survey of Gaussian Convolution Algorithms. *Image Process. Line* **2013**, *3*, 286–310. [CrossRef]

Review

Recent Developments in Instrumentation of Functional Near-Infrared Spectroscopy Systems

Murad Althobaiti and Ibraheem Al-Naib *

Biomedical Engineering Department, College of Engineering, Imam Abdulrahman Bin Faisal University, Dammam 31441, Saudi Arabia; mmalthobaiti@iau.edu.sa
* Correspondence: iaalnaib@iau.edu.sa

Received: 20 August 2020; Accepted: 14 September 2020; Published: 18 September 2020

Abstract: In the last three decades, the development and steady improvement of various optical technologies at the near-infrared region of the electromagnetic spectrum has inspired a large number of scientists around the world to design and develop functional near-infrared spectroscopy (fNIRS) systems for various medical applications. This has been driven further by the availability of new sources and detectors that support very compact and wearable system designs. In this article, we review fNIRS systems from the instrumentation point of view, discussing the associated challenges and state-of-the-art approaches. In the beginning, the fundamentals of fNIRS systems as well as light-tissue interaction at NIR are briefly introduced. After that, we present the basics of NIR systems instrumentation. Next, the recent development of continuous-wave, frequency-domain, and time-domain fNIRS systems are discussed. Finally, we provide a summary of these three modalities and an outlook into the future of fNIRS technology.

Keywords: NIRS technology; spectroscopy; imaging; bioinstrumentation; near-infrared

1. Introduction

The invention of near-infrared spectroscopy (NIRS) enabled many investigations and development in various scientific fields ranging from pure research laboratory experiments to robust industrial procedures for different purposes [1,2]. More interestingly, numerous biomedical studies have been carried out using NIRS systems [3–5]. Among many applications, medical diagnostics, such as functional neuroimaging, cancer diagnosis, rehabilitation, and neurology, have been a drive for numerous investigations and development [3–6]. Starting in the 1990s, a new chapter of NIRS has spawned numerous efforts to develop functional NIRS (fNIRS) systems for different applications [2,7–10]. These efforts followed naturally from the understanding that fNIRS allows functions that are not available by using other techniques. fNIR imaging (fNIRI) and spectroscopy were just some primary examples. More specifically, cerebral blood flow (CBF) and cerebral blood volume (CBV) are indirectly connected with mental activity. Neural activity increases the cerebral metabolic rate of oxygen which consumes glucose and oxygen and releases vasoactive neurotransmitters which lead to vasodilation of arterioles and finally leads to a local increase in CBF and CBV [11]. Therefore, fNIRS is considered one of the main emerging neuroimaging techniques. This allows physicians to view activity within the human brain without the need for quite complicated invasive neurosurgery. The potential of such techniques has been reviewed in several recent papers [6,12–15].

Studying light-tissue interaction at any frequency band is a quite complicated and interesting task at the same time, because the materials of the biological tissues are multilayered, multicomponent, and optically inhomogeneous. It includes reflection, refraction, absorption, and multiple scattering of photons in the tissue as shown in Figure 1. The fact that the absorption by water molecules is lower than oxyhemoglobin and deoxyhemoglobin in the light wavelength range between 650 and

1000 nm enables us to easily estimate the concentration of oxyhemoglobin and deoxyhemoglobin. Nevertheless, strong scattering of light is a characteristic feature of tissue in which near-infrared light propagates in all directions and diffusely illuminates the tissue volume instead of following a narrow path. The absorption and scattering effects at NIR will be discussed in more detail in Section 2.

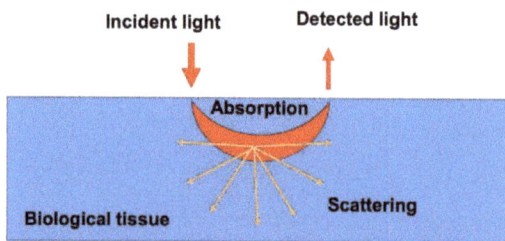

Figure 1. Illustration of the light signal propagation via a biological tissue after it has been partially absorbed and scattered.

In 1993, a NIRS measurement was performed by Hoshi and Tamura [16] by combining five single-channel NIRS instruments. Since then, NIRS instrumentation has been continuously developing and has established its place as a functional brain imaging modality in research use. For instance, a system with 96 sources, 64 detectors, and 3072 measurement channels was recently built [17]. Depending on the area of interest, a number of emitters and detectors are used and separated by a distance of few-to-several centimeters. The acquired raw data is then processed and analyzed using a computer. The estimation of the oxygen saturation of the probed tissue can be evaluated by evaluating the ratio of red-light intensity to the near-infrared light intensity that was re-emitted from the tissue. Therefore, the fraction of oxyhemoglobin measured, for example, at a spot on the brain reflects the local activity at that spot. Furthermore, this method uses a quite low fluence of non-ionizing light radiation. Interestingly, the light wavelengths used have a penetration depth of few centimeters within tissues. Increasing the source–detector separation distance provides a better penetration depth (higher depth sensitivity profile), however, fewer photons will reach the detectors, which results in a low signal-to-noise ratio (SNR). Clearly, there is a trade-off between the light penetration depth and the SNR. A source–detector separation of 3 cm is a reasonable compromise between depth sensitivity and SNR in the brain studies of adults population [18,19] while a source–detector separation of 2 to 2.5 cm is reasonable for the brain of infants population [20,21].

fNIRS instrumentations continually improve, facilitated in part by the availability of compact semiconductor optical photodiodes at the wavelength of interest. Nowadays, commercially compact and wearable systems are also accessible for imaging and spectroscopy. Even with such advancements, fNIRS systems continue to be a subject of practical developments to make them wearable and as compact as possible. The pros and cons of fNIRS have been reviewed in a number of articles [10,22,23]. Ref. [23] specifically discusses the main features of the commercially available fNIRS systems. fNIRS technology is experimentally flexible, silent, and can be easily integrated with positron emission tomography (PET), functional magnetic resonance imaging (fMRI), or electroencephalography (EEG). Nevertheless, fNIRS systems have two main limitations, namely a low spatial resolution of about 1 cm and its ability to get the hemodynamic response at the outer cortex only [23].

Over the last few decades, fNIRS systems have been designed by utilizing different NIRS techniques. These techniques can be categorized into three main types: (i) continuous wave (CW) by measuring the light attenuation using a constant tissue illumination; (ii) frequency-domain (FD) by utilizing the phase delay and attenuation of detected light; and (iii) time-domain (TD) by measuring the shape of short pulses after propagation through tissues. The signal acquired by these techniques is then post-processed using signal processing algorithms. Accordingly, various systems and techniques have been the subject of many informative articles and reviews. Table 1 represents a brief list of those review papers that focused on the fNIRS systems and their applications.

Table 1. Most relevant review papers about CW-, FD-, and TD-fNIRS systems.

Technique	Main Discussed Topics	Year	Ref.
CW-fNIRS	fNIRS major events 1977–2011 Simple point and multi-channel CW fNIRS Main fields of fNIRS applications	2012	[1]
	Technological design aspects (Sources and detectors) Analysis of fNIRI signals	2014	[24]
	Characteristics of key fNIRS and Diffuse Optical Tomography (DOT) technologies	2017	[25]
	fNIRS Most relevant references fNIRS applications	2019	[13]
FD-fNIRS	Fundamental instrumentation design of FD-NIRS	1998	[26]
	Solving the forward model for light propagation in tissue Developments in reconstruction methods	2010	[27]
	A tutorial of the development of Diffuse optical imaging and its applications.	2012	[28]
	Diffuse optical imaging instrumentation	2014	[29]
	DOT image reconstruction algorithms and its clinical applications	2016	[30]
	Basics of FD-NIRS and its application to functional brain studies	2020	[31]
	Recent advances in acquisition and processing speed for several Diffuse Optical Imaging (DOI) modalities.	2020	[32]
TD-fNIRS	Basic features of the TD approach and diffuse optics	2016	[33]
	Theoretical background, instruments, advanced theories and methods	2019	[14]
	Brain monitoring clinical applications	2019	[15]
	Time-gated detection modality	2020	[13]

However, most of the reviews have been focusing on the differences among these three techniques and their applications rather than the instrumentation of such systems. Therefore, the main goal of this paper is to remedy this gap by examining the main differences and similarities between the CW, TD, and FD techniques in building the fNIRS system from an instrumentation point of view. The manuscript is organized as follows: First, the light-tissue interaction at NIR represented by the effects relevant to fNIRS namely the absorption and scattering, and the basics of NIR instrumentation are presented in Section 2. In Sections 3–5, we discuss CW, FD, and TD fNIRS recently developed instrumentations. Section 6 then presents a comparison across these three modalities and finally, the paper is summarized in Section 7.

2. Near-Infrared Systems Instrumentation

As for any optical system at visible or NIR light ranges, instrumentation of NIRS system consists of an (i) emitter device to illuminate a small area of tissue with light at two or more wavelengths, namely red and infrared range, and (ii) a detector device to measure the back-scattered light emerging from the tissue, and (iii) a diffraction grating to enable differentiation and recording the intensity of different wavelengths [24]. Practically, several sources and detectors (that are called optodes) are required and the collective effect is measured.

In order to choose the optimal wavelengths for sources, the effects relevant to fNIRS such as absorption and scattering need to be carefully considered. At NIR wavelengths, the atoms or molecules absorb a part of the light energy. The absorption level is determined based on the molecular composition of tissue, the wavelength of the emitted light, and the thickness of the tissue. Within the NIR window, molecules such as water and lipids are minimal absorbers compared to the iron-containing hemoglobin present within the blood. In this wavelength widow, deoxyhemoglobin (Hb) and oxyhemoglobin (HbO$_2$) absorb light strongly. Figure 2 illustrates these absorption properties of the deoxyhemoglobin (Hb) and oxyhemoglobin (HbO$_2$) as well as the so-called "diagnostic/therapeutic window" where water absorption is at its minimum. Thus, light in this optical window can penetrate deeper in tissue [34–36].

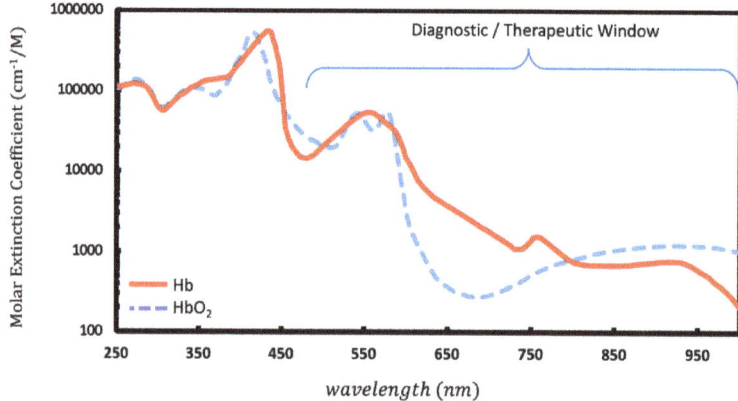

Figure 2. The light absorption of the deoxyhemoglobin (Hb) and oxyhemoglobin (HbO$_2$) of biological tissue (re-drawn from the data taken from Ref. [37]). In the diagnostic window, the absorption level between Hb and HbO$_2$ is notable, and water absorption is at its minimum.

Unlike the absorption, a scattering interaction occurs when light strikes a particle and changes direction. Numerous factors, such as wavelength, particle size, and refractive index of tissue, contribute to the prevalence of scattering. There are two general types of scattering: elastic and inelastic. With elastic scattering, no energy is lost; the light simply changes direction. With inelastic scattering, some energy is lost from the incident light during the interaction, which would mean altered frequency and wavelength. The scattering considered here is the former. Similar to absorption, the intensity of light measured by the detector at a distance in the medium is less than the original intensity of light incident on the tissue described by Beer-Lambert law. For the adult head, due to scattering and absorption, only about one in 10^9 photons that enter the tissue will actually reach the position of a detector located on the surface a few centimeters away from the source. Both absorption and scattering reduce the signal and in actual tissues, they both are present simultaneously [34].

Considering the above effects of light-tissue interaction at NIR, there are a variety of NIR sources currently exist. The incandescent light bulb has often been used in the past while light-emitting diodes (LEDs) are increasingly becoming the main sources in use due to its reliability, low power consumption and its long lifetimes [1,7,13]. At the NIR spectrum range, LEDs are available at different emission wavelengths with output power in the range of mW. Commonly available wavelengths are 660 nm, 670 nm 700 nm, 850 nm, 870 nm and 940 nm. The spectral half-width of LEDs in the 600 nm region is around 20 nm and the widths increase in longer-wavelength materials to around 40 nm for LEDs in the 900 nm region. Nevertheless, wavelength-scanned lasers and frequency combs are used whenever high precision spectroscopy is required [38]. Laser diodes have the advantages of small size, low energy consumption and high output coherent light with output power in the range of mW. At the NIR spectrum range, the GaAs/AlGaAs material (850 nm) and vertical-cavity surface-emitting laser (VCSEL) [39], which range from 750–980 nm, are commonly used.

For typical fNIRS optode separation distances, the intensity of light that penetrates the head and reaches the detector is very small–on the order of only a few mW to pW [40], which is an extremely small value. Therefore, high-sensitivity detectors are crucial in this case. Hence, the choice of detectors includes light-sensitive diodes, photomultiplier tubes (PMTs), semiconductor-based pin, and charge-coupled device (CCD) cameras [41–43] as well as avalanche photodiodes (APDs) [44]. More recently, silicon photomultipliers (SiPMs) have been intensively utilized for fNIRS applications as well [41,44,45]. SiPMs feature major advantages in terms of sensitivity, gain, and speed to acquire the signal [46]. Moreover, they provide a much higher responsivity, three or more orders of magnitude larger than PDs or APDs [41]. The choice of the photodetection devices depends mainly on the intended application and the source emitted wavelengths. For instance, the silicon-based pin is a good choice in the case of the shorter end of the NIR spectrum (400 nm to 1000 nm). In contrast, Germanium and InGaAs based pin photodiodes are suitable for the long-range of the NIR. More specifically, the wavelength of the Germanium pin photodiodes range is from 800 nm to 1600 nm while it is from 1100 nm to 1700 nm for the InGaAs ones. Nevertheless, the responsivity for the silicon-based pin peaks in the range between 800 nm and 900 nm. For silicon and Germanium Avalanche Photodiodes (APD) types, silicon APD has a higher and wider gain (20–400) with minimum dark noise (0.1–1 nA) in comparison to Germanium APD, which has a gain range (50–200) with a dark noise range from 50–500 nA. This makes silicon-based pin a common choice for many fNIRS systems [47,48].

The question that arises here is how many sources and detectors should be used and even more importantly where to locate them. In a recent paper [49], the translation of regions of interest (ROI) to the placement of optodes on a measuring cap has been thoroughly investigated as shown in Figure 3. The authors presented a toolbox in this paper to simplify selecting the right fNIRS optode positions on the scalp. It is based on the overlapping between the simulated photon transport from optodes positioned in 130 positions on the cap and the regions of interest within the brain.

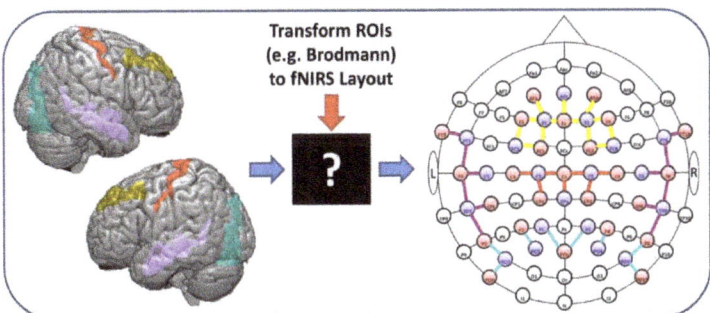

Figure 3. fNIRS cap layout with corresponding color-coded channels. Reproduced from reference [49].

Figure 4 shows a simplistic example of a montage of one source (in orange color) and a total of eight detectors. Hence, we can have up to eight channels to be measured. In case we have more than one source, we need to consider the optical coupling between the light sources and detectors. Actually, it is one of the most important factors affecting the quality of data. The poor coupling may lead to several types of errors in the measurement. These include motion artifacts caused by contact pressure variation, sliding of the probe along the skin, and light leaks. Motion artifacts can lead to signals that are partly or wholly useless. Light leaks may lead to a signal that looks normal but has a lower physiological contrast-to-noise ratio than a signal not affected by a light leakage [50]. Hence, such systems should be checked before taking any measurements against the dark noise.

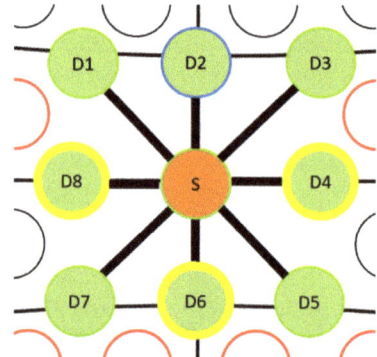

Figure 4. Eight channels montage with one source and eight detectors.

The main property of NIR devices is the maximum possible number of channels. Conventionally, a channel is defined as a possible path between an emitter and a detector. Therefore, the maximum possible number of channels for 8 emitters and 8 detectors system is 8 × 8 = 64 channels. Practically, it is unlikely that the detectors receive a measurable signal from distant emitters, for instance on the other side of an adult head. Hence, detectors are placed within 3–4 cm distance from the corresponding emitters. Depending on the optodes arrangement, a total number of 20–25 channels can be valid using systems with eight emitters and eight detectors [24]. One of the important parameters to consider is the movement artifact due to the mechanical instability of the optodes on the subject. Various designs of optode holders that keep the optodes securely in place and retains a stable contact and pressure against the skin have been proposed [13,25,51]. More importantly, lightweight and comfortable enough caps have been developed to allow some movement for the subject [52].

Several types of systems are currently available for fNIRS measurement techniques: continuous-wave, frequency domain, and time domain. Theoretical light penetration depths and sensitivity profiles are extremely similar for a CW system, a 200 MHz modulated FD system, and a 500 ps pulsed TD system [13]. However, the light signals from FD and TD systems can typically penetrate deeper into the brain than CW systems. Besides, it is feasible with both FD and TD systems to differentiate between the brain and extra-cerebral tissue in superficial regions. Some review papers have already compared these techniques [53,54]. Nevertheless, our paper is concerned with the recent advances of various instrumentation elements of these three systems that have been proposed in the last few years.

3. Continuous Wave fNIRS Instrumentation

The simplest form of tissue spectroscopy methods is the continuous wave technique. It is based on the steady light illumination of tissue and the detection of the transmitted light intensity through the tissue as depicted in Figure 5. In turn, it gives an idea about the relative light attenuation without differentiating the impacts of scattering and absorption. The strongest absorbers present in the blood are the hemoglobin molecules. Hence, valuable information is accessible such as relative changes in blood volume and oxygenation can be obtained. Hence, the relative concentration level can be evaluated with high reliability and contrast from the background. So, it is not a surprise to know that it is currently the most widely used fNIRS technique. The CW technique is very useful as it is very sensitive. Moreover, the sampling rate of less than a second is doable. Furthermore, CW systems can be made to be quite affordable for spectroscopy and imaging as well. In CW systems, the source emits light at the same intensity and the changes in the intensity are measured then by a detector. The light penetration depth increases as the source-detector separation increases, but the measured intensity is less, which leads to a low SNR as illustrated in Figure 5.

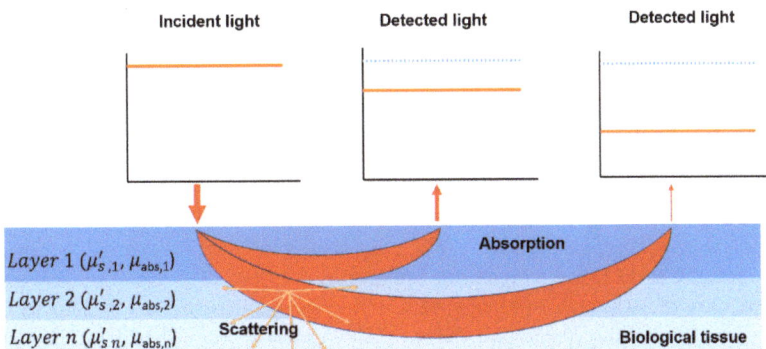

Figure 5. The source emits light at the same intensity and the changes in the intensity are measured then by the detector. The figure illustrates two detected signals successively along with the possible photon paths "banana shape" in different layers with various absorption coefficient and reduced scattering coefficient.

fNIRS systems can be miniaturized quite easily by employing commercially available light sources and detectors. However, achieving wearable fNIRS systems is quite challenging for the fact that high requirements for signal quality and system reliability are required. Most fNIRS systems employ two wavelengths where laser diodes are used as emitters and PMTs or APDs are used as detectors. Figure 6 depicts a block diagram of such a multiwavelength system [55]. Digital gain control is used to equalize all the channels over a 20 dB range. Next, a multiplexer can be used to sample one wavelength at a time. Then, to avoid aliasing at 250 samples/s, the storage capacitor is oversampled by an analog-to-digital converter (ADC) and this gives a temporal resolution of oxygenation measurement of >0.3 s. A modified Beer-Lambert principle is used here and it should be noted that the scattering is considered both homogenous and fixed [24,53,56].

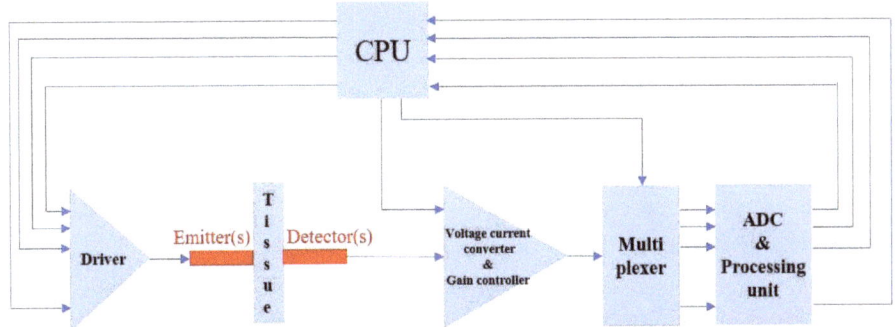

Figure 6. The assembly of a multiwavelength, multisource, multidetector fNIRS system.

Utilizing all eight sources during data collection provides a sampling frequency of 6.25 Hz when the sources are lit sequentially (standard mode). It can be increased by reducing the number of sources used during data collection. Employing only half of the eight sources, for example, results in a sampling frequency of 10.42 Hz-an increase of more than 4 Hz. Detectors are connected to the main unit by optical fibers. Optical fibers consist of a core for transmitting light and a covering to both keep internal light from escaping and external light from entering.

A multiwavelength approach has also been considered via selecting optimum wavelengths over the complete NIR spectrum to find the concentration changes [57]. Although development is being made on devices and methods using this fNIRS multi-spectral approach, they have two drawbacks [58]. It features an increased computational complexity. Moreover, there is a need to reduce the incident power as light with a multiwavelength has higher total power than light with a limited number of wavelengths. Instead, one can sample at the two optimal wavelengths as many times as possible in order to achieve a higher SNR.

In a pioneering work that started in 2015, Von Lühmann et al. suggested a wireless fNIRS for mobile neuroergonomics and Brain-Computer Interface (BCI) Applications [59]. The system uses Time-Division Multiplexing (TDM) of the fNIRS channels. Interestingly, they aimed to have such a system as an open-source instrument. The suggested module offers four dual-wavelength fNIRS channels using 750 and 850 nm with a quite broad emission of 30 to 35 nm. The incoherence and uncollimated characteristics of these sources allow for (i) a stronger strength used for tissue examination, (ii) the optodes to be in direct contact with the scalp as a result of almost no heating of the tissue, and it is (iii) no harm for the eyes. When using TDM, various factors have to be taken into accounts such as inter-channel crosstalk, heating, and battery consumption. More importantly, the SNR is restricted by the width of the used time frames. Figure 7 shows this open-source fNIRS system.

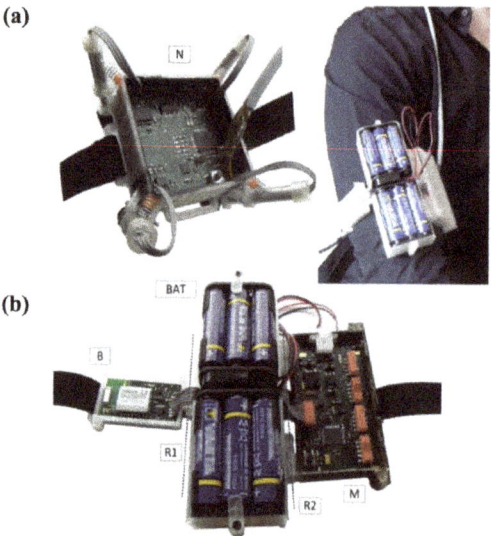

Figure 7. Complete system with (**a**) single 4 channel fNIRS module and (**b**) Bluetooth module. Reproduced from reference [59].

Following that, the original system has been significantly improved and called the ninjaNIRS as shown in Figure 8 [60]. The new system has a very small footprint, scalable that supports up to 128 optodes. They are basically the core of this variable system. The optode itself digitizes the system, the signal and therefore the interface is a purely digital bus. It has a Field Programmable Gate Arrays (FPGA) onboard, at the side of the multi-wavelength led and photodetector. The long-term goal of this study has been to build a high-density fNIRS-EEG-Eye-tracking system that has a long interval to continuously monitor brain activity in real-time during movement, social interaction, and perception whilst being portable, miniaturized, lightweight and wearable.

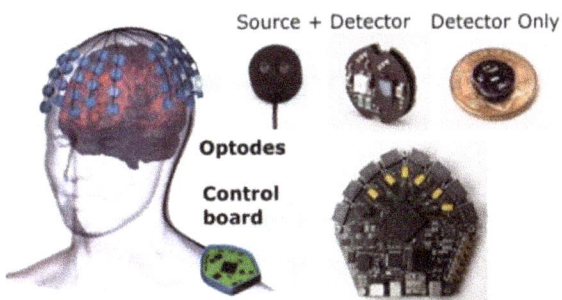

Figure 8. ninjaNIRS optodes and controller. Reproduced from reference [60].

Another very interesting scheme has been proposed in 2017 by Wyser et al. to achieve a wearable and modular fNIRS system with four wavelengths [40] as shown in Figure 9. The scheme features three main characteristics: (i) the ability to measure short-separation (SS) and long-separation (LS) channels, (ii) four wavelengths can be utilized, and (iii) modular optode design that can be put on different brain regions. High modularity is obtained via a miniaturized hardware design of optode modules. It is worth mentioning here that sources and detectors can be individually connected to a central unit. For many fNIRS applications, in particular BCIs, the ability to measure the SS and LS channels can help to detect the desired signal and compensate for unwanted signals. Also, to achieve more robust estimates of concentration changes, four wavelengths are included.

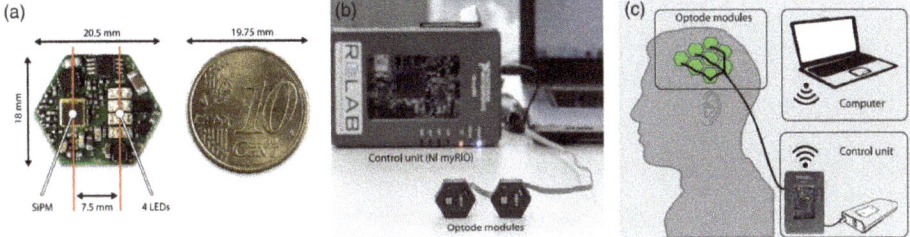

Figure 9. A compact fNIRS instrument. (**a**) The PCB next to a 10-cent euro coin. (**b**) Picture of the fNIRS system with two optode modules. (**c**) Conceptual sketch illustrating the arrangement of the system. Reproduced from reference [40].

The following aspects have been carefully considered during the design process of this system and are summarized in Table 2.

Table 2. Main CW required and achieved parameters [40].

Parameter	Required	Achieved
Signal quality (SNR)	40–60 dB 60 dB	64 dB
Sampling frequency	Above 6 Hz	NS source; ND detector $100\,\text{Hz}/[\text{NS} \times (1 + 0.06 \times \text{ND})]$
Safety	Temperature should not exceed 42 °C Optical power below 10 mW	≤41 °C 1 mW
Usability		Small weight Graphical user interface
Modularity		The number of channels is scalable

The third wearable CW-fNIRS system has been suggested by Chiarelli et al. [61]. It is based on silicon photomultiplier detectors and lock-in amplification with fiber-less and multi-channels. Due to the use of optical fibers in the fNIRS system, mechanical constraints will start being a problem to fNIRS due to the difficulty of stabilizing the optodes onto the scalp to get the required coupling. In fact, to avoid the use of optical fibers, the solution that was proposed in this study is by having direct contact between the sources and detectors from one side and the skin from the other side. The detectors on the scalp will not be allowed to be located as sensitive detectors, such as PMTs, are used. Since PMTs are delicate, bulky, and operate at high voltages, they become impractical in real-life operations. Recently, a solution was employed for fNIRS by using solid-state detectors such as single-photon avalanche diode (SPADs) that feature high sensitivity, although this will make the detector area very small, which is not a favorable solution. In fact, using photodiodes for light detection leads to low sensitivity and a dynamic range of the wearable CW-fNIRS systems.

As illustrated in the block diagram in Figure 10, the designed fNIRS system, named DigiLock consisted of three boards and an FPGA unit. All the necessary components for signal filtering and two sigma-delta converters (TI ADS1298) were used on the ADC board. The LED board implemented 32 time-multiplexed outputs for 16 dual-wavelength LEDs and an adjustable current source. The SiPM board contains an adjustable DC\DC converter for the SiPMs bias generation. Interestingly, during the multiplexing cycle, each combination of LED current source and SiPM bias could be dynamically adjusted for optimal signal acquisition. A single-board computer built around the Xilinx Zynq 7Z020 all programmable system-on-chip (SoC) was based on the FPGA board (MYIR Z-turn). Other useful peripherals were added to the board such as RAM, flash memory, USB, Ethernet, and temperature sensor. Using a tailored hardware description language (HDL) program, all the essential parts needed to execute the lock-in algorithm were implemented within the FPGA. The FPGA handled the elaboration chain implied by the lock-in algorithm and time-sharing synchronization among the LEDs and the ADCs after reading the data from the ADC converters. Employing and using the algorithm, SiPM bias, and automated calibration of each LED current were implemented.

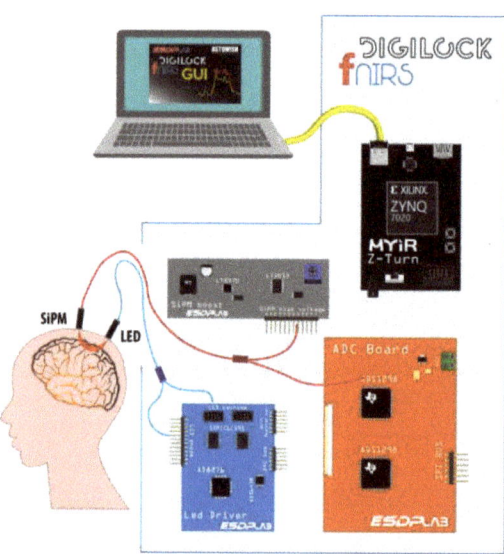

Figure 10. Block diagram of the DigiLock system. Reproduced from reference [61].

The fourth system is a modular, fiberless, and features a flexible-circuit-based wearable fNIRS as shown in Figure 11a,b [62]. It is called Advanced Optical Brain Imaging (AOBI) system. In order to facilitate efficient tessellation, it was designed as a diamond-like shape to cover a surface such as the head surface. This system can be used for quite long time periods with high flexibility of coverage. A single AOBI module has one long-separation channel of 30 mm, four channels with medium-separation of 21.4 mm, and a one-short separation channel of 8 mm.

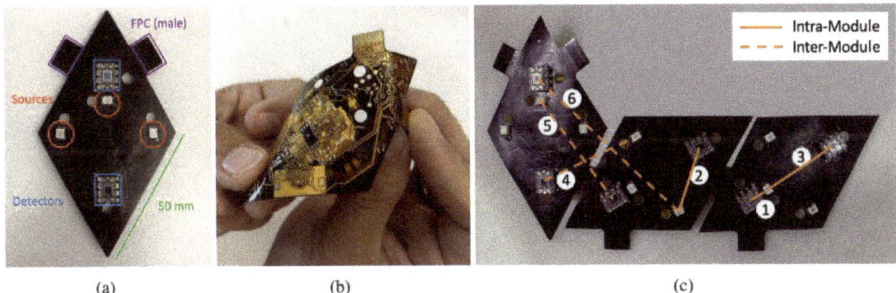

Figure 11. (a) The AOBI module (b) Top view of the flexible circuit board (c) A three-module configuration emphasizing. Reproduced from reference [62].

Figure 11c shows a configuration of three flexible AOBI modules placed over an optical phantom. It results in 54 dual-wavelength channels, 26 of which are below the 40 mm separation used in fNIRS systems. The sampling frequency of all 54 dual-wavelength channels is 33.3 Hz, while a single AOBI module samples at 100 Hz. An SNR of more than 50 dB has been achieved in all intra-module channels, while inter-module channels show an SNR of more than 40 dB up to a 52 mm SD separation.

Thus, this system has features tailored towards full-head coverage and was made following a fibreless, wearable, and modular approach. The flexible-circuit configuration enables the modules to conform and bend to help in enhancing optode-scalp coupling. Moreover, the diamond module shape is well-situated to cover head surfaces. Hence, the main present constraints of the bulky systems with fiber optics can be replaced by much smaller and lighter electrical connections. In the near future, these wearable and low-cost systems will considerably help in acquiring high-density fNIRS measurements. For the preprocessing of the collected fNIRS signals and the filtration of the different types of noise signals (instrumental noise, experimental error, and physiological noise), we refer the reader to Refs. [63–65] for more information in this regard. For advanced postprocessing, feature extractions and classification techniques, the reader is referred to Refs. [66–68].

4. Frequency-Domain fNIRS Instrumentation

Continuous Wave modality is useful to measure light intensity attenuation. However, there are mainly two limitations to the use of CW fNIRS instruments. First, the CW instruments rely on the modified Beer-Lambert principle which assumes a constant scattering degree from all sites of light. The other limitation is the assumption to estimate the light traveled distance, the differential path length (L), where this mode contains no direct information about the time of flight. Hence, it is very difficult to separate the absorption from scattering in a heterogeneous medium using CW systems. Frequency Domain (FD) instruments are the evolution of the CW NIRS instruments. Thirty years ago, FD modality has been recognized as an alternative technique to measure the absorption and scattering coefficients. In the 1990s, the work was about developing the theory and building some prototypes [26,69–73]. Nevertheless, those developments led to what is now the only available commercially FD based instrument by ISS (Table 3). Since 2000, the main works focused on the applications mainly in the area of breast and brain imaging and validating these applications in large clinical studies [74–79].

In FD fNIRS system, NIR light is modulated at a particular radio frequency (RF) usually in the range of a few hundred MHz. The selection of these frequencies is based on the distinct sizes and depths of the imaged object [76]. A high modulation frequency is suitable, for example, for imaging small breast lesions near the surface, while a low modulation frequency is suitable for imaging deeper and larger lesions as illustrated in Figure 12. Ideally, however, all modulation frequencies should be used to obtain the most accurate optical image reconstruction of the imaged object. On the other hand, acquiring NIR measurements with all modulation frequencies is unfavorable in the clinical setting to avoid patient motion. Therefore, one modulation frequency is usually selected for clinical studies based on the depth of the targeted area.

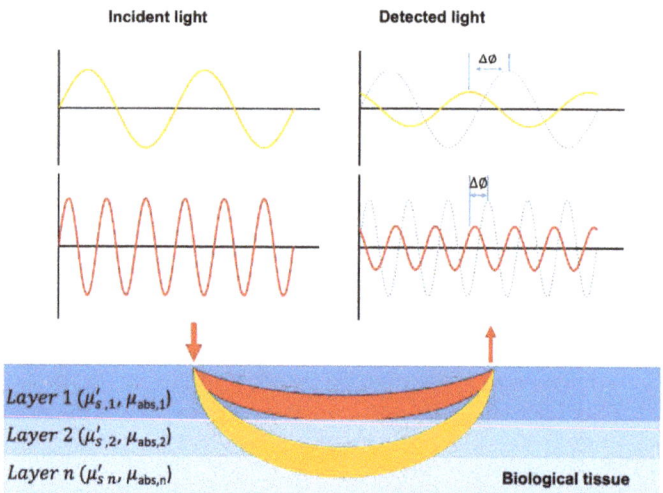

Figure 12. Schematic representation of NIR light penetration for both low and high modulation frequency. The figure illustrates two detected signals successively along with the possible photon paths "banana shape" in different layers with various absorption coefficient and reduced scattering coefficient.

As photon propagates into deeper tissue, its phase shift measurements quantify the degree of the scattered photon in tissue, therefore, tissue scattering parameter is no longer assumed. Both the phase and the amplitude intensity of the attenuated NIRS light can be extracted from the measurements of the FD system. Figure 13 illustrates the principle of dual-phase lock detection [80]. Primarily, the system consists of two mixers, two low-pass filters, and 90^0 phase-shifter. The first mixer mixes the signal and the reference, where the first lowpass filter ensures the output is a DC signal, $S_1 = 0.5$ A $\cos(\phi)$. In the same manner, the 90^0 shifted reference signal is mixed with the original signal where the second lowpass filter ensures the output is a DC signal, $S_2 = 0.5$ A $\sin(\phi)$. From S_1 and S_2, the amplitude-phase unit extracts both the amplitude and phase based on:

$$A = 2\sqrt{S_1^2 + S_2^2} \tag{1}$$

$$\phi = \arctan\frac{S_2}{S_1} \tag{2}$$

If the NIR probe consists of "N" number wavelengths sources, "S" number of sources, and "D" number of detectors, and since both the amplitude and phase at each source-detector pair can be extracted, the resulting total number of measurements for each set of measurements has a total of $M = 2 \times (N \times S \times D)$. As stated earlier, when NIR light penetrates inside the human tissue, scattering of NIR light within human tissue dominates the absorption of the light propagation in such tissue. This imposes a significant challenge to FD-NIRS optical tomography with regard to its spatial resolution and localization accuracy. In fact, the spatial resolution of the optical tomography is limited by the signal-to-noise ratio in the order of 20% of the imaging depth [80].

For breast imaging applications and to overcome the spatial resolution challenge, several groups have studied the co-registration of FD-NIRS optical tomography with other high-resolution imaging modalities such as MRI, ultrasound, and mammography [81–84]. In this approach, high spatial resolution images are used to guide the optical functional imaging with high localization accuracy. One research group has investigated the co-registration of mammographic x-ray images and optical breast imaging. The functional information provided by FD-NIRS optical tomography and the anatomical information provided by mammography imaging offers information that neither mammography nor optical imaging is enough single-handedly [83].

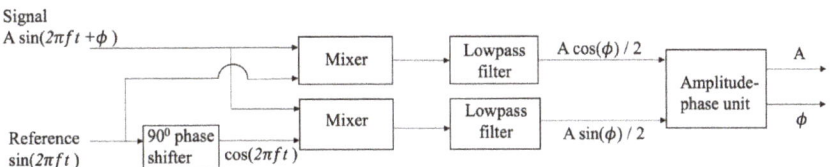

Figure 13. Block diagram of a conventional dual-phase lock detection system.

Once the optical scanning is completed, the optical probe shown in Figure 14a is removed to allow the breast to be compressed for the mammography scanning. The pressure pain associated with mammography scanning is the main disadvantage of this approach. Alternatively, the MRI-guided optical imaging has been investigated to image adipose and fibroglandular breast tissue by another group [82]. In this approach, the optical probe is a circular geometry consists of six laser diodes emit light at two wavelengths 660 and 850 nm modulated at 100 MHz as presented in Figure 14c. For each source illumination, measurements are collected from 15 locations with photomultiplier tube (PMT) detectors. Unlike the mammography and optical breast imaging approach, the MRI and NIR data are acquired simultaneously. The bulk size and the high cost of MRI systems are real challenges for this technique.

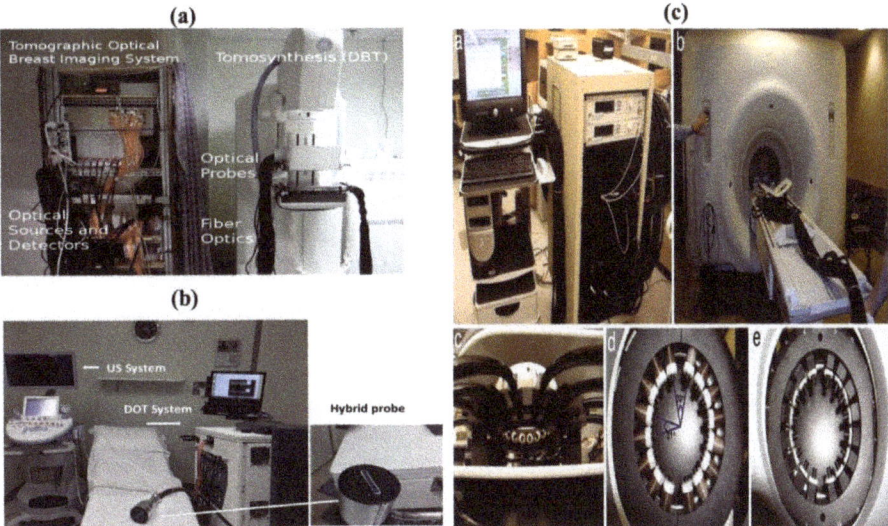

Figure 14. Co-registration of FD-NIRS optical tomography with (**a**) mammography (Reproduced from reference [85]); (**b**) ultrasound (Reproduced from reference [84]); and (**c**) with MRI for breast imaging [82]; Copyright (2006) National Academy of Sciences, U.S.A.

The ultrasound (US)-guided optical imaging approach was also investigated [76,84]. In this approach, the co-registration of the B-scan ultrasound images are utilized to improve the localization of breast tumor, while the optical imaging provides optical absorption information of the tumor vasculature. In the flat surface optical probe geometry illustrated in Figure 14b, the probe consists of 9 source locations and 14 PMT detectors. The NIR light is emitted by four laser diodes of wavelengths 730, 785, 808, and 830 nm modulated at 140 MHz. The B-scan ultrasound probe is located at the center of the optical probe, where it is surrounded by NIR sources and detectors. The ultrasound is considered safe for the patients, its cost is relatively low, and it is a movable system. Figure 14 depicts the three different approaches aimed for breast imaging.

In the last five years, there has been renewed interest in improving the fNIRS technology itself leading to new advantages associated with FD and potentially new applications. Roblyer and his team have recently developed an ultrafast frequency-domain diffuse optics system with a deep neural network (DNN) processing method to measure the optical properties [86]. The DNN is used to replace the time-consuming Levenberg–Marquardt iterative algorithm which was adopted to fit the calibrated amplitude and phase measurements to an analytical forward model. In contrast to the iterative algorithm, the DNN is 3–5 orders of magnitude faster to estimate the optical properties of measured tissue. Therefore, the developed system combined with DNN enables a robust tissue oxygenation monitoring system that can be able to acquire, process, and display absolute concentrations of hemoglobin at an adequate rate to catch the cardiac cycle at the higher speed [86].

In an effort to maximize the penetration depth, Sassaroli et al. has shown that with combinations of sources and detectors using a dual-slope method, phase information can provide deeper sensitivity [87,88]. In this theoretical work, the authors have presented a dual-slope (gradients versus source-detectors separations) method with a requirement of at least two sources and two detectors arranged symmetrically. In comparison to the conventional single-slope method, the dual-slope method has achieved maximal depth sensitivity for all three data types in FD-NIRS (DC, AC, and phase).

In their recent work, Doulgerakis et al. has systematically studied the reconstructed image quality when phase shift measurements incorporated from an FD high-density measurement system [89]. It has been shown that phase information provides not only deeper information sensitivity but also higher effective resolutions than the CW method [89]. This could be potentially very important for fNIRS, where gaining a distance as little as 1 mm or 2 mm means one can reach deeper in the cortex. Both works, by Sassaroli et al. [87] and Doulgerakis et al. [89], showed experimentally that FD systems appear to be sensitive to deeper optical layers. There are different approaches for image reconstruction to recover the optical properties from the FD-NIRS measurements. For instance, a two-step image reconstruction approach was investigated in Refs. [90–92]. Moreover, different regularization techniques such as Tikhonov regularization and Levenberg Marquardt regularization were also studied and the reader is referred to Refs. [93–95] for more information about that.

5. Time-Domain fNIRS Instrumentation

In time-domain fNIRS systems, the tissue is irradiated with picosecond short pulses. At the detector side, detectors that feature very fast responses have to be used in order to record the amplitude of the light pulse as it leaves the tissue as shown in Figure 15. Typically, the received signal is smeared out compared with the original signal as a result of the randomly distributed lengths of photons interacting with different diffusive layers of the tissue and induce various scattering events and form the distribution of time-of-flight (DTOF) of the received photons. Hence, the absorption and scattering properties of the tissue can be assessed using the pulse peak and its time, area, and width. By integrating the temporal profiles, the intensity can be obtained.

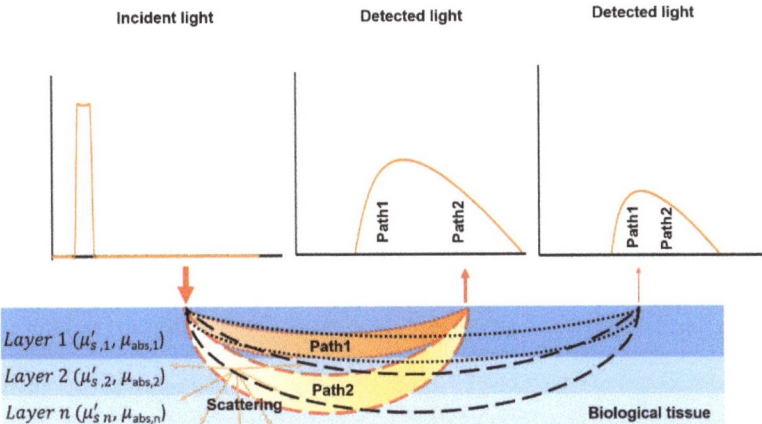

Figure 15. Mechanism of TD NIRS showing the incident light short pulse and two detected signals successively along with the possible photon paths "banana shape" in different layers with various absorption coefficient and reduced scattering coefficient.

Next, the modified Beer-Lambert law can be used to evaluate the absorption variations. Moreover, the mean optical path lengths are calculated from the center of gravity of the temporal profile [96]. The computations of the mean path length and absorption variations are model-independent. Hence, the values of scattering and absorption coefficients (μ_s and μ_a) can be obtained using the nonlinear least-squares method after applying the diffusion equation in reflectance mode into all observed temporal profiles [97]. In turn, absolute concentration levels can be obtained using this technique, which enables time-resolved measurements with any given source-detector distance that in principle can go down to zero. The intracerebral and extracerebral absorption variations are determined then from moments (integral, mean time of flight and variance) of DTOFs [98].

One of the early fNIRS-TD systems is from Hamamatsu Photonics KK, Hamamatsu, Japan [99]. It utilizes three-wavelength with a generated light pulse width of 100 ps, a peak power of 60 mW, an average power of 30 µW, and a pulse rate of 5 MHz as depicted in Figure 16. On the detection side, a PMT was used in photon-counting mode. The received signals are then processed by a TRS circuit. It consists mainly of a time-to-amplitude converter, an ADC, and a histogram memory. Optical fibers are used to illuminate the tissues and to collect the diffuse light accordingly. Two transmitters and two detectors are used and hence two spots can be illuminated at the same time.

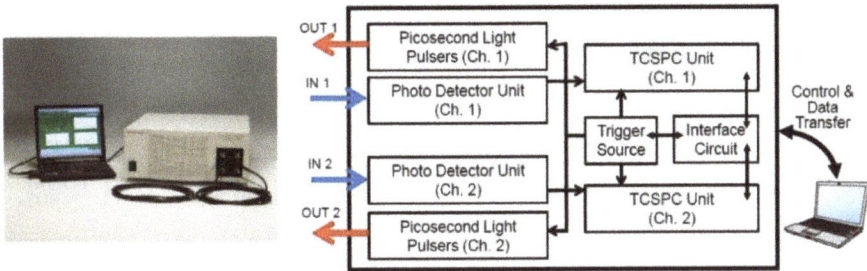

Figure 16. Photograph and schematic diagram of a time-resolved spectroscopy system. Reproduced from reference [14].

There have been some other attempts from academia to build TD-fNIRS systems to assess intracerebral and extracerebral absorption changes as shown in Figure 17 [100]. Indocyanine green (ICG) bolus tracking was utilized for the clinical assessment of brain perfusion at the bedside. Time-correlated single-photon counting (TCSPC) electronics [45] scheme has been utilized for this purpose. A supercontinuum light was generated by fiber lasers with 40 MHz repetition frequency for in-vivo measurements and 80 MHz for in phantom studies. Optical fibers (length = 2 m, NA = 0.22, diameter 400 μm) with used to deliver light to the surface of the phantoms or tissue. A low power level of 20 mW was used. A power density of no more than 2 mW/mm^2 was used at the surface of the skin. A fiber bundle (length 1.5 m, NA = 0.22) was utilized to transfer the photons to the detection system. For in-vivo measurements, the source-detector separation was r = 3 cm and for phantom studies, it was r = 1, 2, 3, and 4 cm. A detector module PML equipped with polychromator (NA = 0.135 and uses 77,414 diffraction grating) and 16-channel PML. Additional losses of photons in the photodetection system were due to the discrepancy between numerical apertures of the detection bundles (0.22) and polychromator (0.135). Nevertheless, lower NA of the using fiber bundles would cause photons losses at the tissue side. Absorption modifications were evaluated from the mean time of flight and variance of its distributions of photons and analyzing the changes in the total number of the received photons measured at 16 wavelengths from the range of 650–850 nm, which replaces earlier technique of measuring at multiple distances with different separations. Phantom, as well as in-vivo measurements, have been carried out for validation.

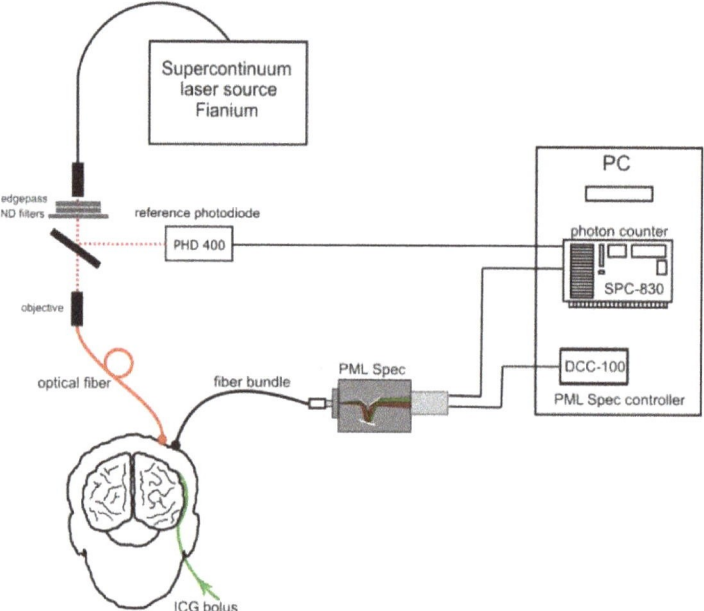

Figure 17. The setup for multiwavelength time-resolved diffuse reflectance measurements. Reprinted with permission from [100] © The Optical Society.

The results of phantom and in-vivo measurements indicated that the optical signal detected at r = 3 cm has a proper quality to assess blood flow in the brain cortex with high precision. The main advantage of this design that it requires a single source-detector separation. A modified algorithm based on the DTOFs acquisition for the single source-detector pair was utilized in this study. The algorithm is based on the assessment of changes of moments of the DTOF's measured at all the wavelengths.

More recently, another TD-fNIRS was built based on four-wave mixing (FWM) laser and fast-gated single-photon avalanche diode and as shown in Figure 18. The laser source was FWM laser delivering light at two wavelengths, namely 710 and 820 nm. The temporal duration of about 25 ps FWHM with a repetition rate of 40 MHz. A variable optical attenuator was used to attenuate the light beam and then a collimator was used to collimate the light into a 400 µm fiber. The sample was then illuminated with the collimated beams. The diffused light was detected using two 1 mm core fibers (NA of 0.37). The separation between the source and the detector was 0.5 or 3 cm. In order to evaluate the concentration of HbO2 and Hb hemoglobin in the in vivo measurements, a filter centered at 710 or 820 nm was utilized to distinguish the two wavelengths. Time-gated detectors (FG-SPAD) modules were used. Another synchronization signal was taken from the laser and split into two parts. The first part was used to feed the FG-SPAD modules in order to trigger the detector. A "stop" signal for both acquiring boards was facilitated by the second part that was fed to the time-correlated single-photon counting (TCSPC) circuit. Accordingly, the "start" signal for the TCSPC was delivered by each FG-SPAD module.

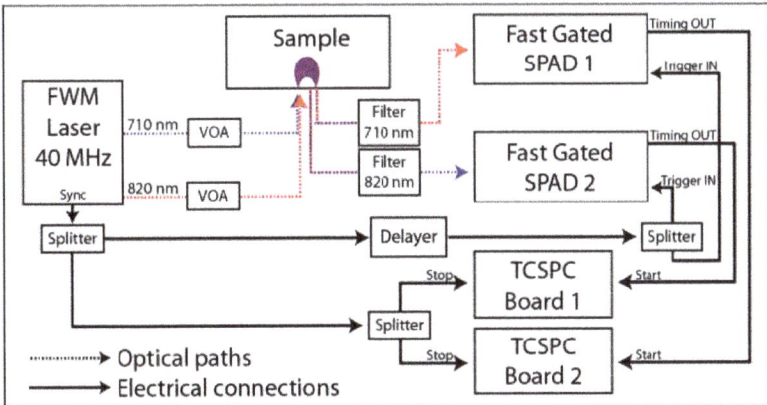

Figure 18. Setup schematics for the in vivo experiment. Reproduced from reference [101].

Phantoms as well as in vivo testing were used for the system characterization. Using the fast-gating technique with a small inter-fiber distance allowed a great increase of the early photons when the space between the source and the detector was reduced. This was a big advantage compared with a non-gated detector as this peak of "early photons" would cause a saturation of the dynamic range, thus decreasing the capability to discriminate a perturbation in depth. This study showed that the gating scheme can enhance the contrast-to-noise-ratio and contrast for the detection of absorption perturbation, irrespective of source-detector distance.

6. CW, FD and TD Comparison and Commercial Systems

FD systems operate by emitting light continuously from a source. That light varies as a sinusoid in intensity with frequencies on the order of megahertz. Detectors measure both the reduction in intensity and the phase shift of the light after it passes through tissue. Combining this information allows a direct measure of absorption and scattering coefficients by assuming that HbO_2 and Hb are the only absorbers that contribute significantly, which eliminates the necessity to define a pathlength for the light. The two main advantages of FD systems are high temporal resolution and absolute quantification of HbO_2 and Hb concentrations. Disadvantages include a relatively large amount of noise within scattering measurements as well as greater complexity and, therefore, cost more than some other NIRS systems. Unlike FD systems, TD systems emit light in short, picosecond-order bursts–or impulses– rather than continuously. These short impulses are broadened to a few nanoseconds, as well as reduced in amplitude, upon transversing biological tissue and the resultant signal is known as either the temporal point spread function (TPSF) or the distribution time-of-flight (DTOF). The broadening of the initial impulse is a consequence of the highly scattering biological tissue; not every photon will follow the same path between source and detector.

By determining the photon's time of flight, path-length can be directly calculated using the speed of light. Like FD systems, TD systems are also able to determine absorption and scattering coefficients. However, TD systems have an even greater overall cost than FD systems. They also require relatively long acquisition times to obtain a reasonable SNR and possess somewhat large dimensions with the need for physical stabilization. An advantage of TD systems over others, though, is the potential for greater spatial resolution as Torricelli and colleagues demonstrated with zero-separation measurements. Similar to FD systems, the light sources of CW systems emit light continuously, as their name implies. Depending upon the specific hardware, the emitted light intensity either has a constant amplitude or varies sinusoidally with frequencies at or below tens of kilohertz. Combining the detected signal intensities with estimates of the differential path-length factor (DPF) allows for calculation–via the MBLL–of relative hemoglobin concentration changes. Primary advantages of CW systems over others include their simplicity, smaller size, and low cost. Hence, CW systems provide the best SNR at sampling frequencies above 1 Hz as well as the potential for the highest sampling rate. However, they also have several disadvantages. CW systems cannot determine absolute quantities of HbO_2 and Hb [24] and cannot distinguish between absorption and scattering. It limits the accuracy as the overall scattering coefficients of the investigated tissues is subject-dependent. Therefore, a single point CW-NIRS only provides variations of hemoglobin concentration. However, measurement of the light attenuation at a number of source/detector separations enables us to estimate the absolute μ_a of the tissue by fitting the measured spatially resolved light attenuation to the solution of the diffusion equation [102]. Last, any change in optode position or amount of pressure against the scalp could significantly alter detected intensities. Finally, Table 3 gives an overview of the commercially available fNIRS systems. Most of those systems are based on the CW technique. Nevertheless, those systems differ when it comes to the number of sources and their types, detectors, and their types, number of channels, the used wavelengths, and sampling rate.

Table 3. fNIRS Commercial Systems.

Company	Product	S	D	C	Source Type	Detector Type	λ (nm)	Sampling Rate (Hz)
Continuous Wave Systems								
Artinis [103]	Brite	10 or 11	7× or 8×	Up to 54	LED	Photodiodes	760, 850	50 or 100
	OxyMon	/	/	Up to 108	Laser	APD	765, 855	50–250
	OctaMon	/	/	8	LED	Photodiode	765, 855	50
	PortaLite	/	/	1 or 3	LED	Photodiode	765, 855	50
Biopac [104]	fNIR100	4	10	16 or 2	LED	Photodiode	730, 850	2
	fNIR2000M Series	/	/	Up to 18	LED	Silicon photodiode	730, 850	5 or 10
Gowerlabs [105]	NTS fNIRS system	6 or 8 or 16	6 or 8 or 16	/	Laser	APD	780, 850	10
	LUMO	3	4	/	/	APD	/	/
Hamamatsu [106]	NIRO-200NX	2	2	2	LED	photodiode	735, 810, 850	/
	ETG-4100	18	16	up to 52	laser	APD	695, 830	10
	ETG-4000	18	16	up to 52	laser	APD	695, 830	10
Hitachi [107]	ETG-7100	/	/	up to 120	laser	APD	695, 830	10
	WOT-100	8	8	16	laser	APD	705, 830	5
NIRX [108]	NIRScout	8–64 (single) or 16–128 (tandem)	4–32 (single) or 8–64 (tandem)	2048	LED or Laser	SP or APD	Laser (2 or 4 λ, 685/780, 808 and 830 nm) or LED (2 λ, 760 nm and 850 nm)	2.5 Hz–62.5 Hz (up to 100 Hz for NIRScoutX and NIRScoutX+)
	NIRSport 2	8–64	8–64	40–60	LED	SP or APD	760, 850	70–240
OBELAB [109]	NIRSIT	24	32	up to 204	laser	/	780, 850	8.138
Rogue Research [110]	Brainsight NIRS	4–16 (24 is possible)	8–32	Up to 72	laser	SP or APD	705, 830	Up to 50
Scenel [111]	Medelopt	16	32	up to 512	LED	/	780, 850	/
Shimadzu [112]	LABNIRS	/	/	up to 142	laser	PT	780, 805, 830	/
	LIGHTNIRS	/	/	10 or 22	laser	APD	780, 805, 830	/
Soterix medical [113]	NIRST	2	4	8	laser	Photodiode	/	8
Spectratech [114]	OEG-17H	6	6	14, 16 or 17	LED	APD	770, 840	0.76, 6.10
TechEn [115]	CW6 System	32	32	/	laser	APD	690, 830	25 Hz
Frequency Domain Systems								
ISS [116]	Imagent	up to 64	up to 32	Up to 512	laser	Photomultiplier tubes	690, 830	Up to 50
Time Domain Systems								
Hamamatsu Photonics [117]	tNIRS-1	2	2	2	Pulsed diode lasers	SiPM	755, 816 and 850	
Politecnico di Milano [38]		2	2	2	Pulsed diode lasers	SiPM	670, 830	

S: Source/D: detector/C: channel; PMT: Photomultiplier tube; APD: avalanche photodiodes; SiPM: Silicon Photomultipliers.

7. Conclusions and Future Perspective

In this paper, we have discussed the recently developed fNIRS systems from an instrumentation point of view. More specifically, the main features, differences, and similarities between the three different modalities (CW, FD, and TD) in building fNIRS systems have been reviewed and discussed. It is evident that the FD modality provides more information than the CW counterpart. Thus, better quantification of optical proprieties of tissue and higher depth sensitivity are possible advantages of the FD technique. However, the complexity of the FD systems and their relatively high cost are clear disadvantages. The recent renewed interest in improving FD fNIRS technology has improved the FD technique to be faster with better resolution and could provide higher sensitivity for imaging deeper tissues. This will pave the way for many potential applications. TD-NIRS systems, on the other hand, have not been as popular as CW-NIRS systems due to their complexity. Nevertheless, we have reviewed a few recent publications that reported TD systems and carried out measurements using phantoms and in-vivo of hemoglobin concentration. Interestingly, with only one channel, it is possible to estimate the optical properties of the tissue from the evaluation of the distribution of the time of flight of photons. Unlike the TD and FD systems, the compact and the simplicity of building CW systems notably allowed this modality to be commercially available for numerous applications. The current development in size and sensitivity of the semiconductor optical detectors will further allow the development of high-density fNIRS systems. With that, more channels could be used for measurements, which ultimately will enhance the quantification of optical properties of tissues.

Author Contributions: Conceptualization, M.A. and I.A.-N.; methodology, M.A. and I.A.-N.; investigation, M.A. and I.A.-N.; resources, M.A. and I.A.-N.; writing—original draft preparation, M.A. and I.A.-N.; writing—review and editing, M.A. and I.A.-N.; visualization, M.A. and I.A.-N.; project administration, M.A.; funding acquisition, M.A. and I.A.-N. All authors have read and agreed to the published version of the manuscript.

Funding: This research was funded by the Deanship of Scientific Research at Imam Abdulrahman Bin Faisal University, Saudi Arabia under project number 2019-013 Eng.

Conflicts of Interest: The authors declare no conflict of interest.

References

1. Ferrari, M.; Quaresima, V. A brief review on the history of human functional near-infrared spectroscopy (fNIRS) development and fields of application. *Neuroimage* **2012**, *63*, 921–935. [CrossRef] [PubMed]
2. Boas, D.A.; Elwell, C.E.; Ferrari, M.; Taga, G. Twenty years of functional near-infrared spectroscopy: Introduction for the special issue. *Neuroimage* **2014**, *85*, 1–5. [CrossRef] [PubMed]
3. Herold, F.; Wiegel, P.; Scholkmann, F.; Müller, N. Applications of Functional Near-Infrared Spectroscopy (fNIRS) Neuroimaging in Exercise–Cognition Science: A Systematic, Methodology-Focused Review. *J. Clin. Med.* **2018**, *7*, 466. [CrossRef]
4. Karthikeyan, P.; Moradi, S.; Ferdinando, H.; Zhao, Z.; Myllylä, T. Optics based label-free techniques and applications in brain monitoring. *Appl. Sci.* **2020**, *10*, 2196. [CrossRef]
5. Soltanlou, M.; Sitnikova, M.A.; Nuerk, H.C.; Dresler, T. Applications of functional near-infrared spectroscopy (fNIRS) in studying cognitive development: The case of mathematics and language. *Front. Psychol.* **2018**, *9*, 277. [CrossRef]
6. Nam, H.S.; Yoo, H. Spectroscopic optical coherence tomography: A review of concepts and biomedical applications. *Appl. Spectrosc. Rev.* **2018**, *53*, 91–111. [CrossRef]
7. Toronov, V.; Franceschini, M.A.; Filiaci, M.; Fantini, S.; Wolf, M.; Michalos, A.; Gratton, E. Near-infrared study of fluctuations in cerebral hemodynamics during rest and motor stimulation: Temporal analysis and spatial mapping. *Med. Phys.* **2000**, *27*, 801–815. [CrossRef]
8. Jue, T.; Masuda, K. *Application of Near Infrared Spectroscopy in Biomedicine*; Springer: Boston, MA, USA, 2013; ISBN 9781461462521.
9. Almulla, L.; Al-Naib, I.; Althobaiti, M. Hemodynamic responses during standing and sitting activities: A study toward fNIRS-BCI. *Biomed. Phys. Eng. Express* **2020**, *6*, 055005. [CrossRef]

10. Fantini, S.; Frederick, B.; Sassaroli, A. Perspective: Prospects of non-invasive sensing of the human brain with diffuse optical imaging. *APL Photonics* **2018**, *3*, 110901. [CrossRef]
11. Raichle, M.E.; Mintun, M.A. Brain work and brain imaging. *Annu. Rev. Neurosci.* **2006**, *29*, 449–476. [CrossRef]
12. Lohani, M.; Payne, B.R.; Strayer, D.L. A review of psychophysiological measures to assess cognitive states in real-world driving. *Front. Hum. Neurosci.* **2019**, *13*, 1–27. [CrossRef] [PubMed]
13. Quaresima, V.; Ferrari, M. A mini-review on functional near-infrared spectroscopy (fNIRS): Where do we stand, and where should we go? *Photonics* **2019**, *6*, 87. [CrossRef]
14. Yamada, Y.; Suzuki, H.; Yamashita, Y. Time-Domain Near-Infrared Spectroscopy and Imaging: A Review. *Appl. Sci.* **2019**, *9*, 1127. [CrossRef]
15. Lange, F.; Tachtsidis, I. Clinical brain monitoring with time domain NIRS: A review and future perspectives. *Appl. Sci.* **2019**, *9*, 1612. [CrossRef]
16. Hoshi, Y.; Tamura, M. Dynamic multichannel near-infrared optical imaging of human brain activity. *J. Appl. Physiol.* **1993**, *75*, 1842–1846. [CrossRef] [PubMed]
17. Zimmermann, B.B.; Deng, B.; Singh, B.; Martino, M.; Selb, J.; Fang, Q.; Sajjadi, A.Y.; Cormier, J.; Moore, R.H.; Kopans, D.B.; et al. Multimodal breast cancer imaging using coregistered dynamic diffuse optical tomography and digital breast tomosynthesis. *J. Biomed. Opt.* **2017**, *22*, 046008. [CrossRef]
18. Strangman, G.E.; Li, Z.; Zhang, Q. Depth Sensitivity and Source-Detector Separations for Near Infrared Spectroscopy Based on the Colin27 Brain Template. *PLoS ONE* **2013**, *8*, e66319. [CrossRef]
19. Li, T.; Gong, H.; Luo, Q. Visualization of light propagation in visible Chinese human head for functional near-infrared spectroscopy. *J. Biomed. Opt.* **2011**, *16*, 045001. [CrossRef]
20. Lloyd-Fox, S.; Blasi, A.; Elwell, C.E. Illuminating the developing brain: The past, present and future of functional near infrared spectroscopy. *Neurosci. Biobehav. Rev.* **2010**, *34*, 269–284. [CrossRef]
21. Taga, G.; Homae, F.; Watanabe, H. Effects of source-detector distance of near infrared spectroscopy on the measurement of the cortical hemodynamic response in infants. *Neuroimage* **2007**, *38*, 452–460. [CrossRef]
22. Yücel, M.A.; Selb, J.J.; Huppert, T.J.; Franceschini, M.A.; Boas, D.A. Functional Near Infrared Spectroscopy: Enabling routine functional brain imaging. *Curr. Opin. Biomed. Eng.* **2017**, *4*, 78–86. [CrossRef]
23. Quaresima, V.; Ferrari, M. Functional Near-Infrared Spectroscopy (fNIRS) for Assessing Cerebral Cortex Function During Human Behavior in Natural/Social Situations: A Concise Review. *Organ. Res. Methods* **2019**, *22*, 46–68. [CrossRef]
24. Scholkmann, F.; Kleiser, S.; Metz, A.J.; Zimmermann, R.; Mata Pavia, J.; Wolf, U.; Wolf, M. A review on continuous wave functional near-infrared spectroscopy and imaging instrumentation and methodology. *Neuroimage* **2014**, *85*, 6–27. [CrossRef] [PubMed]
25. Zhao, H.; Cooper, R.J. Review of recent progress toward a fiberless, whole-scalp diffuse optical tomography system. *Neurophotonics* **2017**, *5*, 1. [CrossRef]
26. Chance, B.; Cope, M.; Gratton, E.; Ramanujam, N.; Tromberg, B. Phase measurement of light absorption and scatter in human tissue. *Rev. Sci. Instrum.* **1998**, *69*, 3457–3481. [CrossRef]
27. Klose, A.D. The forward and inverse problem in tissue optics based on the radiative transfer equation: A brief review. *J. Quant. Spectrosc. Radiat. Transf.* **2010**, *111*, 1852–1853. [CrossRef] [PubMed]
28. O'Sullivan, T.D.; Cerussi, A.E.; Cuccia, D.J.; Tromberg, B.J. Diffuse optical imaging using spatially and temporally modulated light. *J. Biomed. Opt.* **2012**, *17*, 0713111. [CrossRef]
29. Zhang, X. Instrumentation in diffuse optical imaging. *Photonics* **2014**, *1*, 9–32. [CrossRef]
30. Hoshi, Y.; Yamada, Y. Overview of diffuse optical tomography and its clinical applications. *J. Biomed. Opt.* **2016**, *21*, 091312. [CrossRef]
31. Fantini, S.; Sassaroli, A. Frequency-Domain Techniques for Cerebral and Functional Near-Infrared Spectroscopy. *Front. Neurosci.* **2020**, *14*, 1–18. [CrossRef]
32. Applegate, M.B.; Istfan, R.E.; Spink, S.; Tank, A.; Roblyer, D. Recent advances in high speed diffuse optical imaging in biomedicine. *APL Photonics* **2020**, *5*, 040802. [CrossRef]
33. Pifferi, A.; Contini, D.; Mora, A.D.; Farina, A.; Spinelli, L.; Torricelli, A.; Pifferi, A.; Contini, D.; Mora, A.D.; Farina, A.; et al. New frontiers in time-domain diffuse optics, a review. *J. Biomed. Opt.* **2016**, *21*, 091310. [CrossRef] [PubMed]
34. Jacques, S.L. Optical properties of biological tissues: A review. *Phys. Med. Biol.* **2013**, *58*, R37. [CrossRef] [PubMed]

35. Wang, L.V.; Wu, H.-I. *Biomedical Optics*; John Wiley & Sons, Inc.: Hoboken, NJ, USA, 2009; ISBN 9780470177013.
36. Keiser, G. *Biophotonics: Concepts to Applications*; Springer: Singapore, 2016.
37. Tabulated Molar Extinction Coefficient for Hemoglobin in Water. Available online: https://omlc.org/spectra/hemoglobin/summary.html (accessed on 6 September 2020).
38. Buttafava, M.; Martinenghi, E.; Tamborini, D.; Contini, D.; Mora, A.D.; Renna, M.; Torricelli, A.; Pifferi, A.; Zappa, F.; Tosi, A. A Compact Two-Wavelength Time-Domain NIRS System Based on SiPM and Pulsed Diode Lasers. *IEEE Photonics J.* **2017**, *9*, 7800114. [CrossRef]
39. Michalzik, R. *VCSELs-Fundamentals, Technology and Applications of Vertical-Cavity Surface-Emitting Lasers*; Springer: Berlin/Heidelberg, Germany, 2013.
40. Wyser, D.; Lambercy, O.; Scholkmann, F.; Wolf, M.; Gassert, R. Wearable and modular functional near-infrared spectroscopy instrument with multidistance measurements at four wavelengths. *Neurophotonics* **2017**, *4*, 1. [CrossRef]
41. Pagano, R.; Libertino, S.; Sanfilippo, D.; Fallica, G.; Lombardo, S. Improvement of sensitivity in continuous wave near infrared spectroscopy systems by using silicon photomultipliers. *Biomed. Opt. Express* **2016**, *7*, 1183. [CrossRef]
42. Xu, J.; Konijnenburg, M.; Song, S.; Ha, H.; Van Wegberg, R.; Mazzillo, M.; Fallica, G.; Van Hoof, C.; De Raedt, W.; Van Helleputte, N. A 665 μW Silicon Photomultiplier-Based NIRS/EEG/EIT Monitoring ASIC for Wearable Functional Brain Imaging. *IEEE Trans. Biomed. Circuits Syst.* **2018**, *12*, 1267–1277. [CrossRef]
43. Maira, G.; Mazzillo, M.; Libertino, S.; Fallica, G.; Lombardo, S. Crucial aspects for the use of silicon photomultiplier devices in continuous wave functional near-infrared spectroscopy. *Biomed. Opt. Express* **2018**, *9*, 4679. [CrossRef]
44. Adamo, G.; Parisi, A.; Stivala, S.; Tomasino, A.; Agrò, D.; Curcio, L.; Giaconia, G.C.; Busacca, A.; Fallica, G. Silicon photomultipliers signal-to-noise ratio in the continuous wave regime. *IEEE J. Sel. Top. Quantum Electron.* **2014**, *20*, 284–290. [CrossRef]
45. Sanfilippo, D.; Valvo, G.; Mazzillo, M.; Piana, A.; Carbone, B.; Renna, L.; Fallica, P.G.; Agrò, D.; Morsellino, G.; Pinto, M.; et al. Design and development of a fNIRS system prototype based on SiPM detectors. In *Proceedings of the SPIE Photonics West 2014-OPTO: Optoelectronic Devices and Materials*; Kubby, J., Reed, G.T., Eds.; SPIE Press: San Francisco, CA, USA, 2014; Volume 8990, p. 899016.
46. Buzhan, P.; Dolgoshein, B.; Filatov, L.; Ilyin, A.; Kantzerov, V.; Kaplin, V.; Karakash, A.; Kayumov, F.; Klemin, S.; Popova, E.; et al. Silicon photomultiplier and its possible applications. *Nucl. Instrum. Methods Phys. Res. Sect. A Accel. Spectrometers Detect. Assoc. Equip.* **2003**, *504*, 48–52. [CrossRef]
47. Deen, M.J.; Basu, P.K. *Silicon Photonics: Fundamentals and Devices*; Wiley: Hoboken, NJ, USA, 2012.
48. Riesenberg, R.; Wutting, A. Optical detectors. In *Handbook of Biophotonics*; Popp, V.V.J., Tuchin, A., Chiou, S.H.H., Eds.; Wiley: New York, NY, USA, 2011; pp. 297–343.
49. Zimeo Morais, G.A.; Balardin, J.B.; Sato, J.R. FNIRS Optodes' Location Decider (fOLD): A toolbox for probe arrangement guided by brain regions-of-interest. *Sci. Rep.* **2018**, *8*, 1–11. [CrossRef] [PubMed]
50. Noponen, T. *Instrumentation and Methods for Frequency-Domain and Multimodal Near-Infrared Spectroscopy*; Aalto University: Espoo, Finland, 2009; ISBN 9789522482105.
51. Piper, S.K.; Krueger, A.; Koch, S.P.; Mehnert, J.; Habermehl, C.; Steinbrink, J.; Obrig, H.; Schmitz, C.H. A wearable multi-channel fNIRS system for brain imaging in freely moving subjects. *Neuroimage* **2014**, *85*, 64–71. [CrossRef] [PubMed]
52. Von Lühmann, A.; Wabnitz, H.; Sander, T.; Müller, K.R. M3BA: A Mobile, Modular, Multimodal Biosignal Acquisition Architecture for Miniaturized EEG-NIRS-Based Hybrid BCI and Monitoring. *IEEE Trans. Biomed. Eng.* **2017**, *64*, 1199–1210. [CrossRef]
53. Davies, D.J.; Clancy, M.; Lighter, D.; Balanos, G.M.; Lucas, S.J.E.; Dehghani, H.; Su, Z.; Forcione, M.; Belli, A. Frequency-domain vs. continuous-wave near-infrared spectroscopy devices: A comparison of clinically viable monitors in controlled hypoxia. *J. Clin. Monit. Comput.* **2017**, *31*, 967–974. [CrossRef] [PubMed]
54. Diop, M.; Tichauer, K.M.; Elliott, J.T.; Migueis, M.; Lee, T.-Y.; St. Lawrence, K. Comparison of time-resolved and continuous-wave near-infrared techniques for measuring cerebral blood flow in piglets. *J. Biomed. Opt.* **2010**, *15*, 057004. [CrossRef]
55. Chance, B.; Nioka, S.; Zhao, Z.; Imager, W.B. A wearable brain imager. *IEEE Eng. Med. Biol. Mag.* **2007**, *26*, 30–37. [CrossRef]

56. Chiarelli, A.M.; Perpetuini, D.; Filippini, C.; Cardone, D.; Merla, A. Differential pathlength factor in continuous wave functional near-infrared spectroscopy: Reducing hemoglobin's cross talk in high-density recordings. *Neurophotonics* **2019**, *6*, 1. [CrossRef]
57. Nosrati, R.; Vesely, K.; Schweizer, T.A.; Toronov, V. Event-related changes of the prefrontal cortex oxygen delivery and metabolism during driving measured by hyperspectral fNIRS. *Biomed. Opt. Express* **2016**, *7*, 1323. [CrossRef]
58. Yeganeh, H.Z.; Toronov, V.; Elliott, J.T.; Diop, M.; Lee, T.-Y.; Lawrence, K.S. Broadband continuous-wave technique to measure baseline values and changes in the tissue chromophore concentrations. *Biomed. Opt. Express* **2012**, *3*, 2761. [CrossRef]
59. von Lühmann, A.; Herff, C.; Heger, D.; Schultz, T. Toward a wireless open source instrument: Functional near-infrared spectroscopy in mobile neuroergonomics and BCI applications. *Front. Hum. Neurosci.* **2015**, *9*, 1–14. [CrossRef]
60. von Lühmann, A.; Zimmermann, B.B.; Ortega-Martinez, A.; Perkins, N.; Yücel, M.A.; Boas, D.A. Towards Neuroscience in the Everyday World: Progress in wearable fNIRS instrumentation and applications. In Proceedings of the Optics and the Brain 2020, Washington, DC, USA, 20–23 April 2020; p. BM3C.2.
61. Chiarelli, A.M.; Perpetuini, D.; Greco, G.; Mistretta, L.; Rizzo, R.; Vinciguerra, V.; Romeo, M.F.; Merla, A.; Fallica, P.G.; Giaconia, G.C. Wearable, Fiber-less, Multi-Channel System for Continuous Wave Functional Near Infrared Spectroscopy Based on Silicon Photomultipliers Detectors and Lock-In Amplification. In Proceedings of the Annual International Conference of the IEEE Engineering in Medicine and Biology Society, EMBS; IEEE: Berlin, Germany, 2019; pp. 60–66.
62. Vanegas, M.; Dementyev, A.; Mireles, M.; Carp, S.; Fang, Q. A Modular, Fiberless, 3-D Aware, Flexible-circuit-based Wearable fNIRS System. In Proceedings of the Optics and the Brain 2020, Washington, DC, USA, 20–23 April 2020; p. BM3C.3.
63. Cooper, R.J.; Selb, J.; Gagnon, L.; Phillip, D.; Schytz, H.W.; Iversen, H.K.; Ashina, M.; Boas, D.A. A systematic comparison of motion artifact correction techniques for functional near-infrared spectroscopy. *Front. Neurosci.* **2012**, *6*, 147. [CrossRef] [PubMed]
64. Izzetoglu, M.; Devaraj, A.; Bunce, S.; Onaral, B. Motion artifact cancellation in NIR spectroscopy using Wiener filtering. *IEEE Trans. Biomed. Eng.* **2005**, *52*, 934–938. [CrossRef] [PubMed]
65. Huppert, T.J.; Diamond, S.G.; Franceschini, M.A.; Boas, D.A. HomER: A review of time-series analysis methods for near-infrared spectroscopy of the brain. *Appl. Opt.* **2009**, *48*, D280. [CrossRef] [PubMed]
66. Power, S.D.; Chau, T. Automatic single-trial classification of prefrontal hemodynamic activity in an individual with Duchenne muscular dystrophy. *Dev. Neurorehabil.* **2013**, *16*, 67–72. [CrossRef]
67. Power, S.D.; Kushki, A.; Chau, T. Towards a system-paced near-infrared spectroscopy brain-computer interface: Differentiating prefrontal activity due to mental arithmetic and mental singing from the no-control state. *J. Neural Eng.* **2011**, *8*. [CrossRef]
68. Naseer, N.; Hong, K.S. fNIRS-based brain-computer interfaces: A review. *Front. Hum. Neurosci.* **2015**, *9*, 1–15. [CrossRef]
69. Hoshi, Y.; Onoe, H.; Watanabe, Y.; Andersson, J.; Bergström, M.; Lilja, A.; Långstöm, B.; Tamura, M. Non-synchronous behavior of neuronal activity, oxidative metabolism and blood supply during mental tasks in man. *Neurosci. Lett.* **1994**, *172*, 129–133. [CrossRef]
70. Franceschini, M.A.; Moesta, K.T.; Fantini, S.; Gaida, G.; Gratton, E.; Jess, H.; Mantulin, W.W.; Seeber, M.; Schlag, P.M.; Kaschke, M. Frequency-domain techniques enhance optical mammography: Initial clinical results. *Proc. Natl. Acad. Sci. USA* **1997**, *94*, 6468–6473. [CrossRef]
71. Chance, B.; Anday, E.; Nioka, S.; Zhou, S.; Hong, L.; Worden, K.; Li, C.; Murray, T.; Ovetsky, Y.; Pidikiti, D.; et al. A novel method for fast imaging of brain function, non-invasively, with light. *Opt. Express* **1998**, *2*, 411–423. [CrossRef]
72. Liu, H. Low-cost frequency-domain photon migration instrument for tissue spectroscopy, oximetry, and imaging. *Opt. Eng.* **1997**, *36*, 1562. [CrossRef]
73. Yu, G.; Durduran, T.; Furuya, D.; Greenberg, J.H.; Yodh, A.G. Frequency-domain multiplexing system for in vivo diffuse light measurements of rapid cerebral hemodynamics. *Appl. Opt.* **2003**, *42*. [CrossRef] [PubMed]

74. Srinivasan, S.; Pogue, B.W.; Brooksby, B.; Jiang, S.; Dehghani, H.; Kogel, C.; Wells, W.A.; Poplack, S.P.; Paulsen, K.D. Near-infrared characterization of breast tumors in vivo using spectrally-constrained reconstruction. *Technol. Cancer Res. Treat.* **2005**, *4*, 513–526. [CrossRef] [PubMed]
75. Brooksby, B.; Jiang, S.; Dehghani, H.; Pogue, B.W.; Paulsen, K.D.; Weaver, J.; Kogel, C.; Poplack, S.P. Combining near-infrared tomography and magnetic resonance imaging to study in vivo breast tissue: Implementation of a Laplacian-type regularization to incorporate magnetic resonance structure. *J. Biomed. Opt.* **2005**, *10*, 051504. [CrossRef] [PubMed]
76. Zhu, Q.; Xu, C.; Guo, P.; Aquirre, A.; Yuan, B.; Huang, F.; Castilo, D.; Gamelin, J.; Tannenbaum, S.; Kane, M.; et al. Optimal probing of optical contrast of breast lesions of different size located at different depths by US localization. *Technol. Cancer Res. Treat.* **2006**, *5*, 365–380. [CrossRef]
77. Eggebrecht, A.T.; Ferradal, S.L.; Robichaux-Viehoever, A.; Hassanpour, M.S.; Dehghani, H.; Snyder, A.Z.; Hershey, T.; Culver, J.P. Mapping distributed brain function and networks with diffuse optical tomography. *Nat. Photonics* **2014**, *8*, 448–454. [CrossRef]
78. Chitnis, D.; Cooper, R.J.; Dempsey, L.; Powell, S.; Quaggia, S.; Highton, D.; Elwell, C.; Hebden, J.C.; Everdell, N.L. Functional imaging of the human brain using a modular, fibre-less, high-density diffuse optical tomography system. *Biomed. Opt. Express* **2016**, *7*, 4275. [CrossRef]
79. Durduran, T.; Choe, R.; Baker, W.B.; Yodh, A.G. Diffuse optics for tissue monitoring and tomography. *Rep. Prog. Phys.* **2010**, *73*, 76701. [CrossRef]
80. Wang, L.V.; Wu, H. *Biomedical Optics: Principles and Imaging*; John Wiley & Sons: Hoboken, NJ, USA, 2012; ISBN 9780471743040.
81. Zhang, Q.; Brukilacchio, T.J.; Li, A.; Stott, J.J.; Chaves, T.; Hillman, E.; Wu, T.; Chorlton, M.; Rafferty, E.; Moore, R.H.; et al. Coregistered tomographic x-ray and optical breast imaging: Initial results. *J. Biomed. Opt.* **2005**, *10*, 024033. [CrossRef]
82. Brooksby, B.; Pogue, B.W.; Jiang, S.; Dehghani, H.; Srinivasan, S.; Kogel, C.; Tosteson, T.D.; Weaver, J.; Poplack, S.P.; Paulsen, K.D. Imaging breast adipose and fibroglandular tissue molecular signatures by using hybrid MRI-guided near-infrared spectral tomography. *Proc. Natl. Acad. Sci. USA* **2006**, *103*, 8828–8833. [CrossRef]
83. Fang, Q.; Selb, J.; Carp, S.A.; Boverman, G.; Miller, E.L.; Brooks, D.H.; Moore, R.H.; Kopans, D.B.; Boas, D.A. Combined optical and x-ray tomosynthesis breast imaging. *Radiology* **2011**, *258*, 89–97. [CrossRef]
84. Vavadi, H.; Mostafa, A.; Zhou, F.; Uddin, K.M.S.; Althobaiti, M.; Xu, C.; Bansal, R.; Ademuyiwa, F.; Poplack, S.; Zhu, Q. Compact ultrasound-guided diffuse optical tomography system for breast cancer imaging. *J. Biomed. Opt.* **2018**, *24*, 021203. [CrossRef] [PubMed]
85. Fang, Q.; Carp, S.A.; Selb, J.; Boverman, G.; Zhang, Q.; Kopans, D.B.; Moore, R.H.; Miller, E.L.; Brooks, D.H.; Boas, D.A. Combined optical imaging and mammography of the healthy breast: Optical contrast derived from breast structure and compression. *IEEE Trans. Med. Imaging* **2009**, *28*, 30–42. [CrossRef] [PubMed]
86. Zhao, Y.; Applegate, M.B.; Istfan, R.; Pande, A.; Roblyer, D. Quantitative real-time pulse oximetry with ultrafast frequency-domain diffuse optics and deep neural network processing. *Biomed. Opt. Express* **2018**, *9*, 5997. [CrossRef] [PubMed]
87. Sassaroli, A.; Blaney, G.; Fantini, S. Dual-slope method for enhanced depth sensitivity in diffuse optical spectroscopy. *J. Opt. Soc. Am. A* **2019**, *36*, 1743. [CrossRef] [PubMed]
88. Blaney, G.; Sassaroli, A.; Pham, T.; Fernandez, C.; Fantini, S. Phase dual-slopes in frequency-domain near-infrared spectroscopy for enhanced sensitivity to brain tissue: First applications to human subjects. *J. Biophotonics* **2020**, *13*. [CrossRef]
89. Doulgerakis, M.; Eggebrecht, A.T.; Dehghani, H. High-density functional diffuse optical tomography based on frequency-domain measurements improves image quality and spatial resolution. *Neurophotonics* **2019**, *6*, 035007. [CrossRef]
90. Althobaiti, M.; Vavadi, H.; Zhu, Q. An automated preprocessing method for diffuse optical tomography to improve breast cancer diagnosis. *Technol. Cancer Res. Treat.* **2018**, *17*, 1–12. [CrossRef]
91. Xu, S.; Shihab Uddin, K.M.; Zhu, Q. Improving DOT reconstruction with a Born iterative method and US-guided sparse regularization. *Biomed. Opt. Express* **2019**, *10*, 2528. [CrossRef]
92. Tavakoli, B.; Zhu, Q. Two-step reconstruction method using global optimization and conjugate gradient for ultrasound-guided diffuse optical tomography. *J. Biomed. Opt.* **2013**, *18*, 016006. [CrossRef]

93. Katamreddy, S.H.; Yalavarthy, P.K. Model-resolution based regularization improves near infrared diffuse optical tomography. *J. Opt. Soc. Am. A* **2012**, *29*, 649. [CrossRef]
94. Guven, M.; Yazici, B.; Intes, X.; Chance, B. Diffuse optical tomography with a priori anatomical information. *Phys. Med. Biol.* **2005**, *50*, 2837–2858. [CrossRef] [PubMed]
95. Althobaiti, M.; Vavadi, H.; Zhu, Q. Diffuse optical tomography reconstruction method using ultrasound images as prior for regularization matrix. *J. Biomed. Opt.* **2017**, *22*, 026002. [CrossRef] [PubMed]
96. Zhang, H.; Miwa, M.; Urakami, T.; Yamashita, Y.; Tsuchiya, Y. Simple subtraction method for determining the mean path length traveled by photons in turbid media. *Jpn. J. Appl. Physics Part 1 Regul. Pap. Short Notes Rev. Pap.* **1998**, *37*, 700–704. [CrossRef]
97. Ijichi, S.; Kusaka, T.; Isobe, K.; Okubo, K.; Kawada, K.; Namba, M.; Okada, H.; Nishida, T.; Imai, T.; Itoh, S. Developmental changes of optical properties in neonates determined by near-infrared time-resolved spectroscopy. *Pediatr. Res.* **2005**, *58*, 568–573. [CrossRef] [PubMed]
98. Liebert, A.; Wabnitz, H.; Steinbrink, J.; Obrig, H.; Möller, M.; Macdonald, R.; Villringer, A.; Rinneberg, H. Time-resolved multidistance near-infrared spectroscopy of the adult head: Intracerebral and extracerebral absorption changes from moments of distribution of times of flight of photons. *Appl. Opt.* **2004**, *43*, 3037–3047. [CrossRef]
99. Oda, M.; Yamashita, Y.; Nakano, T.; Suzuki, A.; Shimizu, K.; Hirano, I.; Shimomura, F.; Ohmae, E.; Suzuki, T.; Tsuchiya, Y. *Near-Infrared Time-Resolved Spectroscopy System for Tissue Oxygenation Monitor*; EOS/SPIE European Biomedical Optics Week: Amsterdam, The Netherlands, 2000.
100. Gerega, A.; Milej, D.; Weigl, W.; Kacprzak, M.; Liebert, A. Multiwavelength time-resolved near-infrared spectroscopy of the adult head: Assessment of intracerebral and extracerebral absorption changes. *Biomed. Opt. Express* **2018**, *9*, 2974. [CrossRef]
101. Di Sieno, L.; Dalla Mora, A.; Torricelli, A.; Spinelli, L.; Re, R.; Pifferi, A.; Contini, D.; Di Sieno, L.; Mora, A.D.; Torricelli, A.; et al. A versatile setup for time-resolved functional near infrared spectroscopy based on fast-gated single-photon avalanche diode and on four-wave mixing laser. *Appl. Sci.* **2019**, *9*, 2366. [CrossRef]
102. Matcher, S.J.; Kirkpatrick, P.J.; Nahid, K.; Cope, M.; Delpy, D.T. Absolute quantification methods in tissue near-infrared spectroscopy. In *Proceedings of the Optical Tomography, Photon Migration, and Spectroscopy of Tissue and Model Media: Theory, Human Studies, and Instrumentation*; Chance, B., Alfano, R.R., Eds.; SPIE: San Jose, CA, USA, 1995; Volume 2389, pp. 486–495.
103. Artinis. Available online: https://www.artinis.com (accessed on 11 July 2020).
104. Biopac. Available online: https://www.biopac.com (accessed on 11 July 2020).
105. Gowerlabs. Available online: https://www.gowerlabs.co.uk (accessed on 11 July 2020).
106. Hamamats. Available online: https://www.hamamatsu.com/eu/en/product/type/C10448/index.html (accessed on 11 July 2020).
107. Hitachi. Available online: https://www.hitachi-hightech.com/global/ (accessed on 11 July 2020).
108. NIRX. Available online: https://nirx.net (accessed on 11 July 2020).
109. OBELAB. Available online: http://obelab.com/index.php (accessed on 11 July 2020).
110. Rogue Research. Available online: https://www.rogue-research.com/nirs/ (accessed on 11 July 2020).
111. Seenel. Available online: https://seenel-imaging.com (accessed on 11 July 2020).
112. Shimadzu. Available online: https://www.shimadzu.com/an/index.html (accessed on 11 July 2020).
113. Soterix Medical. Available online: https://soterixmedical.com (accessed on 11 July 2020).
114. Spectratech. Available online: https://www.spectratech.co.jp/En/productEn.html (accessed on 11 July 2020).
115. TechEn. Available online: https://www.nirsoptix.com/CW6.html (accessed on 11 July 2020).
116. ISS. Available online: http://iss.com/biomedical/instruments/imagent.html (accessed on 11 July 2020).
117. Fujisaka, S.I.; Ozaki, T.; Suzuki, T.; Kamada, T.; Kitazawa, K.; Nishizawa, M.; Takahashi, A.; Suzuki, S. A clinical tissue oximeter using nir time-resolved spectroscopy. In *Proceedings of the Advances in Experimental Medicine and Biology*; Springer: New York, NY, USA, 2016; Volume 876, pp. 427–433.

© 2020 by the authors. Licensee MDPI, Basel, Switzerland. This article is an open access article distributed under the terms and conditions of the Creative Commons Attribution (CC BY) license (http://creativecommons.org/licenses/by/4.0/).

Article

Characterization and Optimal Design of Silicon-Rich Nitride Nonlinear Waveguides for 2 µm Wavelength Band

Zhihua Tu [1], Daru Chen [2], Hao Hu [3], Shiming Gao [1,4] and Xiaowei Guan [3,*]

[1] Centre for Optical and Electromagnetic Research, State Key Laboratory of Modern Optical Instrumentation, International Research Centre for Advanced Photonics, Zhejiang University, Hangzhou 310058, China; tuzhihua@zju.edu.cn (Z.T.); gaosm@zju.edu.cn (S.G.)
[2] Hangzhou Institute of Advanced Studies, Zhejiang Normal University, Hangzhou 311231, China; daru@zjnu.cn
[3] DTU Fotonik, Department of Photonics Engineering, Technical University of Denmark, 2800 Kgs. Lyngby, Denmark; huhao@fotonik.dtu.dk
[4] Ningbo Research Institute, Zhejiang University, Ningbo 315100, China
* Correspondence: xgua@fotonik.dtu.dk

Received: 9 October 2020; Accepted: 12 November 2020; Published: 15 November 2020

Abstract: Optical communication using the 2 µm wavelength band is attracting growing attention for the sake of mitigating the information 'capacity crunch' on the way, where on-chip nonlinear waveguides can play vital roles. Here, silicon-rich nitride (SRN) ridge waveguides with different widths and rib heights are fabricated and measured. Linear characterizations show a loss of ~2 dB/cm of the SRN ridge waveguides and four-wave mixing (FWM) experiments with a continuous wave (CW) pump reveal a nonlinear refractive index of ~1.13×10^{-18} m^2/W of the SRN material around the wavelength 1950 nm. With the extracted parameters, dimensions of the SRN ridge waveguides are optimally designed for improved nonlinear performances for the 2 µm band, i.e., a maximal nonlinear figure of merit (i.e., the ratio of nonlinearity to loss) of 0.0804 W^{-1} or a super-broad FWM bandwidth of 518 nm. Our results and design method open up new possibilities for achieving high-performance on-chip nonlinear waveguides for long-wavelength optical communications.

Keywords: four-wave mixing; nonlinear figure of merit; silicon-rich nitride; ridge waveguide; conversion bandwidth

1. Introduction

The capacity of the current optical communication systems operating at the mainstream wavelengths, i.e., around 1310 nm and 1550 nm is approaching a shortage due to the explosive growth in information being transmitted. In order to exploit new frequency resources for telecommunications, transmitting signals at mid-infrared (MIR) wavelengths has been put on the agenda [1–6], among which the 2 µm wavelength band (1900 to 2100 nm) attracts special attention as signals in this band can be significantly amplified by thulium-doped fiber amplifiers (TDFAs) [7] and glass fiber in this band is still transparent [8]. Indeed, besides the amplifier, many other functional devices necessary for the 2-µm-band optical communication system have been also commercially available or laboratory developed including the laser [9,10], modulator [11], low-loss hollow-core fiber [12], wavelength-division multiplexer [13,14], detector [15,16], etc. Meanwhile, nonlinear devices like the wavelength converter are also indispensable and, in order to leverage the manufacturing scalability for such devices, it is preferred to use the on-chip nonlinear waveguides with the fabrication processes compatible with the mature silicon-based complementary metal-oxide-semiconductor (CMOS) technology.

As to the CMOS-compatible nonlinear waveguides, silicon (Si) waveguides and stoichiometric silicon nitride (Si_3N_4) waveguides are currently the workhorses at the telecommunication wavelengths [17–21]. However, Si still suffers from the two-photon absorption (TPA) below the wavelength 2200 nm [22], despite that linear Si waveguides can possess a low loss in the 2 µm band [23] and the Si nonlinear waveguides have been used for parametric conversion around 2 µm [24]. Indeed, key nonlinear devices like the parametric amplifier [25] or the frequency comb generator [26] have been demonstrated using the Si nonlinear waveguides in the MIR wavelengths but just over the wavelength 2200 nm. For Si_3N_4, its low linear refractive index (RI) inevitably gives a large footprint of any Si_3N_4-based photonic device and the low nonlinear RI usually produces a poor energy efficiency of devices based on the Si_3N_4 nonlinear waveguides. Moreover, in spite of demonstrations of high-performance Si_3N_4 waveguides fabricated by using some advanced processes [27], strain issues are still present in Si_3N_4 film, especially in the thick Si_3N_4 film, which are yet necessary for achieving compact devices at long wavelengths like 2 µm and for waveguides with proper dispersion. Besides, titanium dioxide (TiO_2) waveguides [4] and silicon germanium (SiGe) waveguides [28] were also demonstrated to transmit optical signals at 2 µm, but, up to now, no reports can be found for them being used for nonlinear processes in this band.

Recently, silicon-rich nitride (SRN) has emerged as a promising candidate for the on-chip nonlinear optics [29–39] since SRN possesses larger linear and nonlinear RIs from a Si excess compared to the stoichiometric Si_3N_4 and, more importantly, the SRN film can have a substantially reduced stress [40]. While the SRN waveguides have been applied for many nonlinear processes like the octave-spanning supercontinuum generation (SCG) [31] and the frequency comb generation (FCG) [32], they were almost operating at the near infrared wavelengths. A SRN waveguide was very recently used to transmit optical signals at 2 µm and the wavelength conversion was also demonstrated with a pulse pump in the 2 µm band, but the waveguide was quite thin (300 nm) and the dispersion was not optimized [39]. Here, we design and fabricate SRN ridge waveguides and perform linear and nonlinear characterizations of them, which show a propagation loss of ~2 dB/cm and a moderate nonlinearity of ~4.62 W^{-1} m^{-1} of the fabricated SRN waveguides, corresponding to a nonlinear RI of ~1.13 × 10^{-18} m^2/W of the SRN material at the wavelength 1.95 µm. Then the SRN nonlinear ridge waveguides are optimized by adjusting the width and the rib height to achieve the maximal nonlinear figure of merit (FOM) and to achieve a broad four-wave mixing (FWM) bandwidth of 518 nm.

2. Structure and Fabrication

Figure 1a shows the schematic structure of the present SRN ridge waveguide with the total height, rib height and width denoted as H, h, and w, respectively. A thick SRN layer was needed for tightly confining light in the SRN core and obtaining a proper dispersion profile. Here, we used a low-pressure chemical vapor deposition (LPCVD) machine to deposit an 810 nm SRN film on 2.5 µm thermal silicon dioxide (SiO_2). The SRN film was first deposited by a reaction between dichlorsilane (SiH_2Cl_2 at 80 sccm) and ammonia (NH_3 at 20 sccm) at 830 °C and a pressure of 120 mTorr, and then annealed for three hours at 1150 °C. After annealing, the Si excess was measured to be ~18.7% and the SRN film was condensed to 800 nm. A thick resist (CSAR, ~1.1 µm) was then spun and patterned by the electron beam lithography. The patterns were then transferred to the SRN film by using an inductive coupled plasma machine at −10 °C with flows of the etching gases SF_6/Ar/CF_4/CH_4 being 4/10/4/2 sccm. SRN ridge waveguides with various widths and rib heights were fabricated with a fixed H of 800 nm. After residual resist stripping, the ridge waveguides were finalized with air as the upper cladding and the dimensions were measured, as shown by an exemplified scanning electron microscopy (SEM) image in Figure 1b. RIs at near infrared wavelengths of the deposited and annealed SRN film were measured by an ellipsometry method and fitted with the Sellmeier formula to extract the RIs around 2 µm wavelengths. Here, the RI was calculated to be 2.123 at 1950 nm. Figure 1c shows the simulated mode profile of the fundamental transverse-electric (TE_0) mode for a SRN ridge waveguide with

$w = 1.32$ μm and $h = 640$ nm at the wavelength 1950 nm. The power confinement ratio in the SRN core was calculated to be 89.2%, indicating strong confinement of the SRN ridge waveguide.

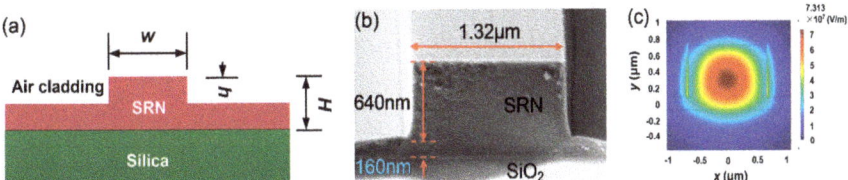

Figure 1. The fabricated silicon-rich nitride (SRN) ridge waveguide. (**a**) Schematic diagram of the structure. (**b**) Scanning electron microscopy image. (**c**) Simulated fundamental transverse-electric (TE$_0$) mode profile at 1950 nm of the waveguide with a width of 1.32 μm and a rib height of 640 nm.

3. Linear and Nonlinear Characterizations

3.1. Linear Characterizations

Linear transmission spectra of the fabricated SRN ridge waveguides with lengths of 5.5 mm, 13 mm and 21 mm were measured from 1860 nm to 1980 nm. Then the cut-back method was used to extract the propagation losses and the coupling losses to a tapered lensed fiber for waveguides with different w and h. In order to reduce the accidental error, six identical waveguides were fabricated and measured for each waveguide dimension configuration. To serve as an example, Figure 2a,b exhibits the propagation and coupling losses with respect to the wavelength for the fabricated SRN ridge waveguides with $h = 380$ nm and $w = 950$ nm, and $h = 640$ nm and $w = 1320$ nm, respectively. Due to the Fabry–Perot interference effect between the two facets of a waveguide, the nonperfect uniformity of the six identical waveguides and the measurement inaccuracy, there were statistical errors for the extracted losses, as shown in the figures. Nevertheless, the propagation losses of the fabricated SRN ridge waveguides were around 2 dB/cm in the investigated wavelength range. Meanwhile, the coupling losses were around 4 dB/facet. Results on the other waveguide dimensions are summarized in Table 1 in the end of this section.

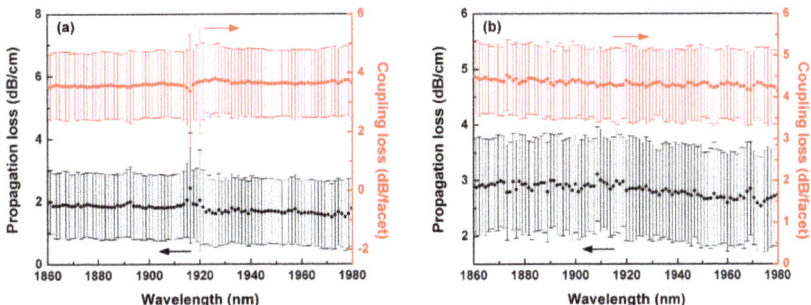

Figure 2. The measured propagation loss and coupling loss with respect to the wavelength for the fabricated silicon-rich nitride ridge waveguides with (**a**) $h = 380$ nm and $w = 950$ nm and (**b**) $h = 640$ nm and $w = 1320$.

The propagation loss α of the fabricated SRN ridge waveguide includes linear absorption loss α_1 and scattering loss α_2. α_1 can be derived from the imaginary part of the effective RI, $N_{eff,r}$ of the SRN ridge waveguide by considering the absorption of the SRN material and expressed as [41]

$$\alpha_1 = \frac{2\omega}{c}\text{Im}(N_{eff,r}) \quad (1)$$

where ω is the optical angular frequency and c is the speed of light in vacuum. The scattering loss α_2 depends on the sidewall roughness, half the waveguide width $w/2$, RI steps from the SRN core to the surroundings, and scaling factor of a ridge waveguide to a channel waveguide (i.e., $h = H$). The sidewall roughness is characterized by the correlation length L_c and the mean squared error σ^2 deviated from the sidewall surface. Thus, the scattering loss α_2 can be expressed as Equation (2) [42]

$$\alpha_2 = 4.343 \frac{\sigma^2}{\sqrt{2} k_0 (w/2)^4 N_{eff,r}} g \cdot f \cdot s \qquad (2)$$

where the unit of α_2 is dB/m, k_0 is the wave vector in vacuum, and function g is completely determined by the dimensions of the SRN channel waveguide and f is relevant to L_c and RI steps. Details for calculating g and f can be found in Payne and Lacey's work for analyzing a planar waveguide's scattering loss using the exponential autocorrelation function [43]. The scaling factor s can be calculated as [42]

$$s = \frac{\delta N_{eff,r}/\delta d}{\delta N_{eff,c}/\delta d} \qquad (3)$$

where $N_{eff,c}$ is the effective RI of the SRN channel waveguide and d is half the waveguide width $w/2$.

A larger h and a smaller w mean more SRN being etched and more light fields overlapping the rough sidewalls and thus a larger scattering loss. In contrast, for the SRN ridge waveguide with a small h and a large w, almost all the light is confined inside the SRN core and, thus, α is dominantly determined by the absorption loss α_1. Here, we reasonably assume α_1 to be 1.7 dB/cm considering it is the minimal propagation loss value we have extracted for the waveguides with different dimensions and from the ridge waveguide with a very shallow etch, i.e., $h = 230$ nm and $w = 840$ nm. Thus, with Equation (2), we can use the measured propagation losses of the SRN ridge waveguides with different dimensions to fit L_c and σ^2. The fitted values of (5, 11.3) nm for (σ, L_c) were found to be able to provide nice fittings of the calculated losses using Equation (2) to the measured propagation losses, as shown in Figure 3, and meanwhile consistent with the values ($\sigma = 5$ nm, $L_c = 45$ nm) reported in other literature where the sidewall roughness of a fabricated SRN waveguide was characterized [30]. Thus, it is reasonable to use Equation (2) to predict propagation losses of the SRN ridge waveguide with any w and h.

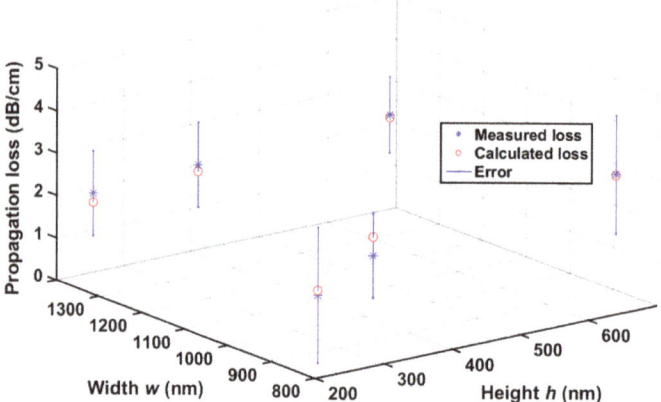

Figure 3. The measured and calculated propagation losses of the fabricated silicon-rich nitride ridge waveguides with different dimensions.

3.2. Nonlinear Characterizations

With the experimental setup shown in Figure 4, FWM experiments were implemented to characterize the nonlinear parameter γ of the fabricated SRN ridge waveguides with different dimensions. The pump signal was provided by a commercial continuous-wave (CW) laser (AdValue Photonics AP-CW1) with a fixed wavelength of 1950.1 nm. The probe signal was generated by a homemade CW laser having a thulium doped fiber as the gain media pumped at 793 nm. The wavelength of our homemade CW laser was set at 1953 nm. High power isolators (HP-ISOs) were used to protect the CW lasers and the pump and signal lights were combined by a 9:1 coupler. The combination of a fiber polarization beam splitter (PBS) and a fiber polarization controller (PC) was implemented to guarantee both the pump light and the signal light were TE-polarized when injected into the waveguides by a tapered lensed fiber. The output light was collected by another lensed fiber and the FWM spectra were recorded by an optical spectrum analyzer (OSA, Yokogawa AQ6375B).

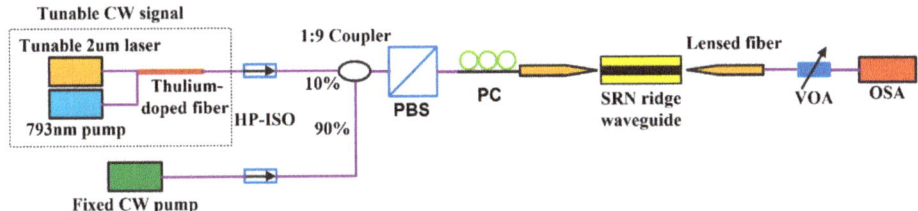

Figure 4. The experimental setup for four-wave mixing (FWM) experiments in the fabricated silicon-rich nitride (SRN) ridge waveguides. CW: continuous-wave; HP-ISO: high power isolator; PBS: polarization beam splitter; PC: polarization controller; VOA: variable optical attenuator; OSA: optical spectrum analyzer.

The idler signal with $\omega_I = 2\omega_P - \omega_S$ can be generated in the SRN ridge waveguide via the degenerate FWM, where $\omega_I, \omega_P, \omega_S$ are the frequencies of the idler, pump and signal lights, respectively. Since the deposited SRN possessing a moderate excess of Si, the TPA and free-carrier absorption (FCA) can be neglected in the 2 μm band. Considering the linear loss, self-phase modulation, cross-phase modulation and FWM, the interaction relationships among these three lights can be described by the following full-vectorial coupling equations [41]

$$\frac{\partial A_P}{\partial z} = -\frac{1}{2}\alpha_P A_P + j\gamma_P\left(|A_P|^2 + 2|A_S|^2 + 2|A_I|^2\right)A_P + 2j\gamma_P A_S A_I A_P^* \exp(j\Delta\beta z) \quad (4)$$

$$\frac{\partial A_S}{\partial z} = -\frac{1}{2}\alpha_S A_S + j\gamma_S\left(|A_S|^2 + 2|A_P|^2 + 2|A_I|^2\right)A_S + j\gamma_S A_P^2 A_I^* \exp(-j\Delta\beta z) \quad (5)$$

$$\frac{\partial A_I}{\partial z} = -\frac{1}{2}\alpha_I A_S + j\gamma_I\left(|A_I|^2 + 2|A_P|^2 + 2|A_S|^2\right)A_I + j\gamma_I A_P^2 A_S^* \exp(-j\Delta\beta z) \quad (6)$$

where A_m ($m = P, S, I$) is the field amplitude of the pump, signal, or idler, α_m is the linear loss, $\Delta\beta$ is the linear phase mismatch, and γ_m is the nonlinear parameter, respectively. When the interacting waves are all in the same wavelength band, γ_m can be expressed as Equation (7) [44]

$$\gamma_m = \frac{\omega_m}{cA_{eff}(\omega_m)}\overline{n}_2(\omega_m) \quad (7)$$

where A_{eff} is the effective mode area and $\bar{n}_2$ is the effective nonlinear RI. They can be calculated as [45]

$$A_{eff}(\omega_m) = \frac{\left|\iint \text{Re}\left[\vec{e}(x,y,\omega_m) \times \vec{h}^*(x,y,\omega_m)\right] \cdot \hat{e}_z dxdy\right|^2}{\iint \left\{\text{Re}\left[\vec{e}(x,y,\omega_m) \times \vec{h}^*(x,y,\omega_m)\right] \cdot \hat{e}_z\right\}^2 dxdy} \tag{8}$$

$$\bar{n}_2(\omega_m) = \frac{\varepsilon_0}{\mu_0} \frac{\iint n_2(x,y,\omega_m) n^2(x,y,\omega_m) \left|\vec{e}(x,y,\omega_m)\right|^2 \left|\vec{e}^*(x,y,\omega_m)\right|^2 dxdy}{\iint \left\{\text{Re}\left[\vec{e}(x,y,\omega_m) \times \vec{h}^*(x,y,\omega_m)\right] \cdot \hat{e}_z\right\}^2 dxdy} \tag{9}$$

where $n(x,y,\omega_m)$ and $n_2(x,y,\omega_m)$ are the linear and nonlinear RIs of the material at position (x,y) at the frequency ω_m, respectively. $\vec{e}(x,y,\omega_m)$ and $\vec{h}(x,y,\omega_m)$ are the electric and magnetic field distributions on the waveguide transverse plane. $\hat{e}_z$ is the unit vector along the propagation direction. ε_0 and μ_0 are the dielectric constant and the permeability constant, respectively. Finally, the conversion efficiency η (in the unit of dB) is defined as the ratio of the output idler power to the output signal power, that is

$$\eta(\text{dB}) = -10\lg\frac{|A_I(L)|^2}{|A_S(L)|^2} \tag{10}$$

where L is the physical length of the waveguide.

We have measured the output FWM spectra of 21-mm-long SRN ridge waveguides with different waveguide dimensions under various coupled pump powers and extracted the conversion efficiencies (CEs), when the signal wavelength and the incident signal power were fixed at 1953 nm and 10 mW, respectively. For example, Figure 5a,b shows the measured FWM spectra of the waveguide with h = 380 nm and w = 950 nm under a coupled pump power of 61.5 mW, and h = 640 nm and w = 1320 under a coupled pump power of 52.4 mW, respectively. CEs of -53.2 dB and -51.1 dB can be extracted from the spectra. It should be noted that there was already some idler light generated in the incident fiber. We have normalized the output idler powers to that coming from the fiber and found little difference on the CEs between the normalizations before and after. Figure 5c,d shows the measured and normalized CEs with respect to the coupled pump power for the waveguides with h = 380 nm and w = 950 nm, and h = 640 nm and w = 1320 nm, respectively. By solving the equations from (4) to (10), we can fit the measured CEs versus the coupled pump power and the nonlinear parameter γ can be extracted to be 2.79 $W^{-1} m^{-1}$ for the waveguide with h = 380 nm and w = 950 nm and 4.62 $W^{-1} m^{-1}$ for the waveguide with h = 640 nm and w = 1320, respectively. With the extracted γ value, the nonlinear index n_2 of the SRN material can also be calculated by using Equation (7). Here, we assume the whole nonlinear effect was contributed by the SRN material since the silica surroundings have a nonlinear RI orders lower than that of the SRN material [45].

We have summarized the linear and nonlinear properties of the fabricated SRN ridge waveguides with different waveguide dimensions in Table 1. While there were variations for these measured or extracted values, the fabricated waveguides generally exhibited a linear loss of ~ 2 dB/cm and a nonlinear index of ~1.13×10^{-18} m^2/W. These values are indeed consistent with that from literatures where the SRN material and waveguide were measured at 1550 nm [30].

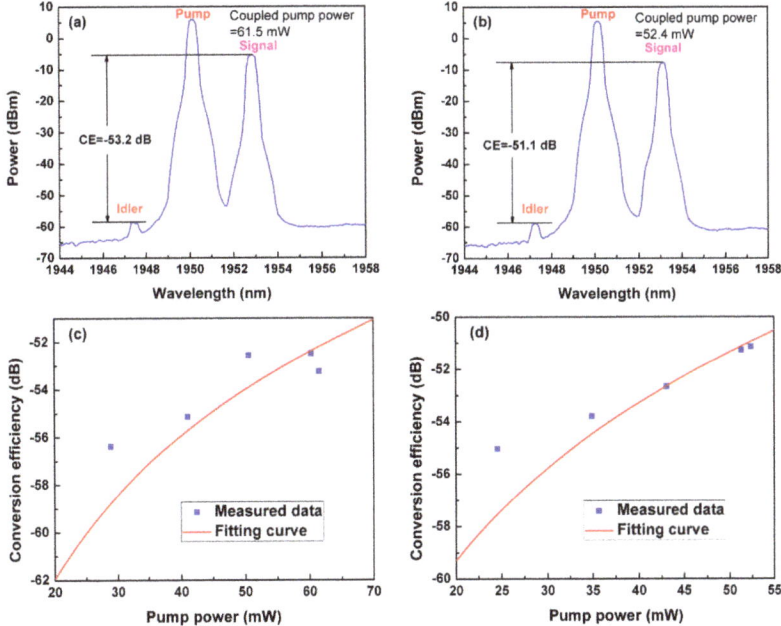

Figure 5. Measured output four-wave mixing spectra (**a**,**b**) and the wavelength conversion efficiencies with respect to the coupled pump powers (**c**,**d**) of the fabricated silicon-rich nitride ridge waveguides with a length of 21 mm. (**a**) and (**c**) are for the waveguide with h = 380 nm and w = 950 nm. (**b**) and (**d**) are for the waveguide with h = 640 nm and w = 1320 nm.

Table 1. Summary of the linear and nonlinear properties of the fabricated silicon-rich nitride ridge waveguides with various rib heights (h) and widths (w) at 1950 nm.

	Measured α (dB/cm)	Calculated α (dB/cm)	γ (W^{-1} m^{-1})	n_2 (m^2/W)
h = 230 nm, w = 840 nm	1.7 ± 1.6	1.8	3.74	1.77 × 10^{-18}
h = 230 nm, w = 1360 nm	2.1 ± 1.0	1.9	2.90	1.30 × 10^{-18}
h = 380 nm, w = 950 nm	1.7 ± 1.0	2.1	2.79	0.87 × 10^{-18}
h = 380 nm, w = 1350 nm	2.3 ± 1.0	2.1	2.49	0.80 × 10^{-18}
h = 640 nm, w = 800 nm	3.3 ± 1.4	3.3	5.98	1.24 × 10^{-18}
h = 640 nm, w = 1320 nm	2.7 ± 0.9	2.6	4.62	1.13 × 10^{-18}

4. Optimal Design for the Maximal Nonlinear Energy Efficiency

For waveguides without nonlinear losses, we can use a figure of merit (FOM) to evaluate the nonlinear energy efficiency of the waveguide, this is the ratio of the waveguide nonlinear parameter to the linear loss, expressed as [46]

$$\text{FOM} = \frac{\gamma}{\alpha} \quad (11)$$

For the proposed SRN ridge waveguide, both γ and α are dependent on the width and rib height. While a deeper etching and a moderately smaller width can give a stronger confinement and hence a larger nonlinearity, this also leaves more light overlapping the rough sidewalls and therefore gives a larger linear propagation loss. Thus, the FOM can be used to fairly evaluate the overall nonlinear efficiency of the SRN ridge waveguide with different dimensions. We have calculated the nonlinear parameter, linear loss and FOM of the SRN ridge waveguides for many combinations of w and h by knowing the linear and nonlinear RIs of SRN and the fabrication quality (i.e., L_c and σ) and shown them in Figure 6a–c, respectively. As expectations, γ and α generally have the similar varying trend as the waveguide width or etch depth changes. This finally yielded a maximal value of 0.0804 W^{-1} of the FOM at the rib height h = 700 nm and width w = 1100 nm. Here, γ an α were calculated to be 5.50 W^{-1} m^{-1} and 2.97 dB/cm, respectively.

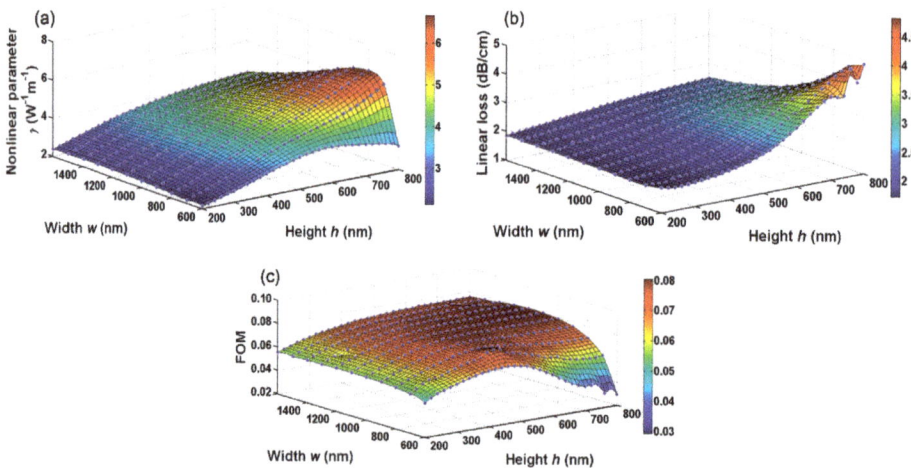

Figure 6. The calculated waveguide nonlinear parameter γ (**a**), linear loss α (**b**) and figure of merit (FOM) (**c**) of a silicon-rich nitride ridge waveguide with respect to the width w and rib height h. Here, the total height H is fixed at 800 nm.

We have also calculated the nonlinear performances of the FWM wavelength conversion for the designed SRN ridge waveguide with the maximal FOM (h = 700 nm, w = 1100 nm). Figure 7a shows the CEs as the waveguide length increases, when the input pump wavelength/power and input signal wavelength/power were set to be 1950 nm/500 mW and 1951 nm/1 mW, respectively. Here, CE is defined as the ratio of the output idler power to the input signal power. The nonlinear effect for generating new idler photons dominated the power of the idler for a short waveguide while for a long waveguide, the linear loss took charge. Thus, there was an optimal length for maximal CE and, here in Figure 7a, a length of 14.6 mm was obtained for a maximal CE of −36.2 dB. It is worth noting that, to obtain a higher CE, one can further increase the pump power or use a resonating structure to enhance the nonlinear interactions between light and the SRN ridge waveguide. Figure 7b shows the calculated wavelength dependence of the CE for the optimized waveguide with a length of 14.6 mm at a fixed pump wavelength of 1950 nm and an adjusted signal wavelength. The 3 dB conversion bandwidth was found to be 94 nm.

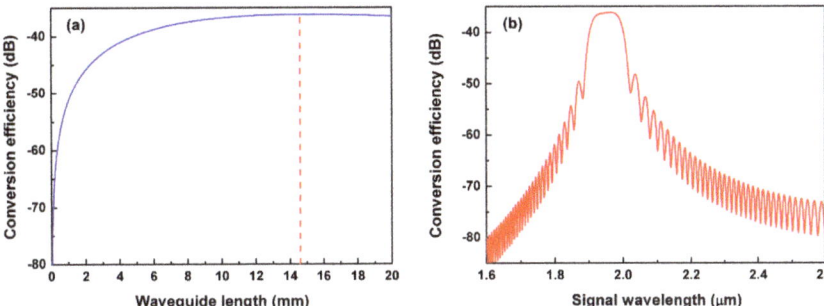

Figure 7. The calculated four-wave mixing conversion efficiency for the designed silicon-rich nitride ridge waveguide (h = 700 nm, w = 1100 nm) with a maximal figure of merit (FOM). (**a**) Dependence on the waveguide length. Here, difference of the pump and signal wavelengths is 1 nm. (**b**) Dependence on the signal wavelength. Here, the waveguide length is 14.6 mm and the pump wavelength is 1950 nm.

5. Optimal Design for a Superbroad FWM Conversion Bandwidth

Although the optimally designed waveguide above exhibits a maximal nonlinear efficiency, the bandwidth is still very limited due to a dispersion which is not small enough (164.3 ps/nm/km) at 1950 nm as shown by the calculated wavelength dependence of the dispersion in Figure 8a (green solid line). Figure 8a also shows the calculated wavelength-dependent dispersion curves for the SRN ridge waveguides with some other dimensions. The dispersion was found to be close to 0 (−6.0 ps/nm/km) at 1950 nm for the waveguide with h = 625 nm and w = 1050 nm. Such a low dispersion is expected to give a broad nonlinear bandwidth. For such an SRN ridge waveguide, the waveguide nonlinear parameter and the linear loss were calculated to be 5.24 $W^{-1}m^{-1}$ and 2.88 dB/cm, respectively, thus producing a FOM of 0.0790 W^{-1}. This FOM is only a little compromised compared with the maximal one (0.0804 W^{-1}). Furthermore, the calculated FWM CE was calculated to exhibit a maximal value of −36.22 dB for a waveguide length of 15 mm, when the input pump wavelength/power and input signal wavelength/power were set to be 1950 nm/500 mW and 1951 nm/1 mW, respectively. Figure 8b shows the calculated wavelength dependence of the FWM CE for the designed SRN ridge waveguide and the 3 dB conversion bandwidth was found to be as broad as 518 nm thanks to the small waveguide dispersion around 2 μm. Besides, one can also find that it may be not a good idea to etch the SRN layer through, i.e., h = 800 nm, see the solid brown curve in Figure 8a, for a low dispersion in the 2 μm band, which is nevertheless mostly implemented in the 1550 nm band.

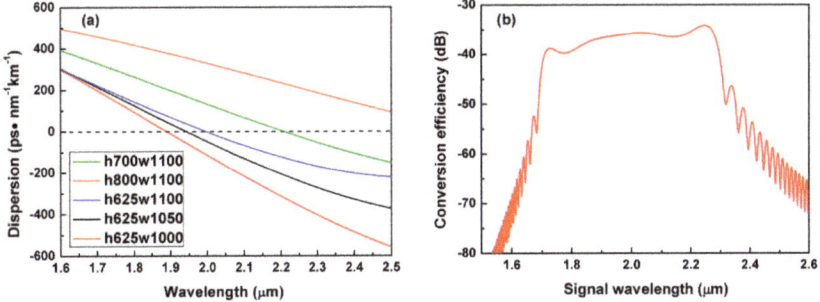

Figure 8. (**a**) Calculated wavelength dependence of the dispersion for the silicon-rich nitride ridge waveguides with various waveguide dimensions. (**b**) Calculated four-wave mixing conversion efficiency with respect to the signal wavelength for a SRN ridge waveguide (h = 625 nm, w = 1050 nm) with the zero-dispersion wavelength at 1950 nm.

6. Discussion and Conclusions

We have proposed and fabricated the silicon-rich nitride ridge waveguides and characterized their linear and nonlinear performances at 2 μm wavelengths. SRN ridge waveguides with different rib heights and widths were fabricated, exhibiting a linear loss around 2 dB/cm. Four-wave mixing experiments with a CW pump at 1950 nm were performed in the fabricated SRN ridge waveguides and revealed a waveguide nonlinear parameter of ~3-6 $W^{-1}m^{-1}$ around 2 μm. With the measured and extracted parameters which characterize the fabrication quality and the material property, optimal design of the SRN ridge waveguides was carried out and a maximal FOM of 0.0804 W^{-1} was found among ridge waveguides with various dimension configurations at 1950 nm. Meanwhile, the ridge waveguide could also be designed to achieve the FWM conversion with a superbroad bandwidth (518 nm) and little compromise of the nonlinear FOM. First, these results show that a stripe silicon nitride waveguide may not be the best choice under some fabrication quality and material properties when moving the operating wavelengths from the conventional telecommunication band to longer wavelengths like the 2 μm band. Second, although the nonlinear conversion efficiency (−51.1 dB) achieved in our fabricated SRN ridge waveguide was much smaller than that of a silicon waveguide (−10 dB [24]) and an SRN waveguide (−18 dB [39]) pumped with pulses, it is indeed the first time, to the best of our knowledge, the nonlinear properties of a silicon nitride waveguide in the 2 μm spectral window have been revealed using a CW pump. Last, the proposed design method for optimal nonlinear performances, in terms of either FOM or the bandwidth, is expected to open new avenues for achieving better on-chip nonlinear waveguides for applications involving longer wavelengths like the 2 μm band.

Author Contributions: Conceptualization, design and fabrication, Z.T. and X.G.; Measurement, Z.T., and D.C.; Data analysis, Z.T., S.G. and X.G.; Writing—original draft preparation, Z.T.; writing—review and editing, D.C., H.H., S.G. and X.G.; All authors have read and agreed to the published version of the manuscript.

Funding: National Natural Science Foundation of China (Grant No. 61875172 and 61475138), Zhejiang Provincial Natural Science Foundation of China (Grant No. LD19F050001 and LY19F050014), Det Frie Forskningsråd (Danish Council for Independent Research) (DFF-7107-00242) and Villum Fonden (Villum Foundation) (023112, 00023316 and 15401).

Conflicts of Interest: The authors declare no conflict of interest.

References

1. Gunning, F.; Corbett, B. Time to Open the 2-μm Window? *Opt. Photonics News* **2019**, *30*, 42–47. [CrossRef]
2. Su, Y.; Wang, W.; Hu, X.; Hu, H.; Huang, X.; Wang, Y.; Si, J.; Xie, X.; Han, B.; Feng, H.; et al. 10 Gbps DPSK transmission over free-space link in the mid-infrared. *Opt. Express* **2018**, *26*, 34515–34528. [CrossRef] [PubMed]
3. Tuan, T.H.; Nishiharaguchi, N.; Suzuki, T.; Ohishi, Y. Mid-infrared transmission by a tellurite hollow core optical fiber. *Opt. Express* **2019**, *27*, 30576–30588. [CrossRef]
4. Lamy, M.; Finot, C.; Fatome, J.; Arocas, J.; Weeber, J.-C.; Hammani, K. Demonstration of High-Speed Optical Transmission at 2 μm in Titanium Dioxide Waveguides. *Appl. Sci.* **2017**, *7*, 631. [CrossRef]
5. Lee, E.; Luo, J.; Sun, B.; Ramalingam, V.L.; Yu, X.; Wang, Q.; Yu, F.; Knight, J.C. 45W 2 μm nanosecond pulse delivery using antiresonant hollow-core fiber. In Proceedings of the Conference on Lasers and Electro-Optics: Science and Innovations, San Jose, CA, USA, 13–18 May 2018.
6. Kong, D.; Liu, Y.; Ren, Z.; Jung, Y.; Pu, M.; Yvind, K.; Galili, M.; Oxenløwe, L.K.; Richardson, D.J.; Hu, H. Generation and Coherent Detection of 2-μm-band WDM-QPSK Signals by On-chip Spectral Translation. In Proceedings of the Optical Fiber Communication Conference (OFC), San Diego, CA, USA, 8–12 March 2020.
7. Li, Z.; Heidt, A.M.; Daniel, J.M.O.; Jung, Y.; Alam, S.U.; Richardson, D.J. Thulium-doped fiber amplifier for optical communications at 2 μm. *Opt. Express* **2013**, *21*, 9289–9297. [CrossRef]
8. Agrawal, G.P. *Nonlinear Fiber Optics*, 3rd ed.; Academic: Boston, MA, USA, 2001.

9. Geng, J.; Wang, Q.; Jiang, S. 2μm fiber laser sources and their applications. In Proceedings of the NanOphotonics and Macrophotonics for Spacel Envgineering + Applicationments V, San Diego, CA, USA, 13 September 2011.
10. Yang, C.-A.; Xie, S.-W.; Zhang, Y.; Shang, J.-M.; Huang, S.-S.; Yuan, Y.; Shao, F.-H.; Zhang, Y.; Xu, Y.; Niu, Z.-C. High-power, high-spectral-purity GaSb-based laterally coupled distributed feedback lasers with metal gratings emitting at 2 μm. *Appl. Phys. Lett.* **2019**, *114*, 021102. [CrossRef]
11. Cao, W.; Hagan, D.; Thomson, D.J.; Nedeljkovic, M.; Littlejohns, C.G.; Knights, A.; Alam, S.-U.; Wang, J.; Gardes, F.Y.; Zhang, W.; et al. High-speed silicon modulators for the 2 μm wavelength band. *Optica* **2018**, *5*, 1055–1062. [CrossRef]
12. Richardson, D.J. New optical fibres for high-capacity optical communications. *Philos. Trans. R. Soc. A Math. Phys. Eng. Sci.* **2016**, *374*, 20140441. [CrossRef]
13. Ryckeboer, E.; Gassenq, A.; Muneeb, M.; Hattasan, N.; Pathak, S.; Cerutti, L.; Rodriguez, J.; Tournié, E.; Bogaerts, W.; Baets, R.; et al. Silicon-on-insulator spectrometers with integrated GaInAsSb photodiodes for wide-band spectroscopy from 1510 to 2300 nm. *Opt. Express* **2013**, *21*, 6101–6108. [CrossRef]
14. Liu, Y.; Li, Z.; Li, D.; Yao, Y.; Du, J.; He, Z.; Xu, K. Thermo-Optic Tunable Silicon Arrayed Waveguide Grating at 2-μm Wavelength Band. *IEEE Photonics J.* **2020**, *12*, 1–8. [CrossRef]
15. Wang, R.; Vasiliev, A.; Muneeb, M.; Malik, A.; Sprengel, S.; Boehm, G.; Amann, M.-C.; Šimonytė, I.; Vizbaras, A.; Vizbaras, K.; et al. III-V-on-silicon photonic integrated circuits for spectroscopic sensing in the 2–4 μm wavelength range. *Sensors* **2017**, *17*, 1788. [CrossRef] [PubMed]
16. Xu, S.; Wang, W.; Huang, Y.-C.; Dong, Y.; Masudy-Panah, S.; Wang, H.; Gong, X.; Yeo, Y.-C. High-speed photo detection at two-micron-wavelength: Technology enablement by GeSn/Ge multiple-quantum-well photodiode on 300 mm Si substrate. *Opt. Express* **2019**, *27*, 5798–5813. [CrossRef]
17. Leuthold, J.; Koos, C.; Freude, W. Nonlinear silicon photonics. *Nat. Photonics* **2010**, *4*, 535–544. [CrossRef]
18. Espinola, R.L.; Dadap, J.I.; Osgood, J.R.M.; McNab, S.J.; Vlasov, Y.A. C-band wavelength conversion in silicon photonic wire waveguides. *Opt. Express* **2005**, *13*, 4341–4349. [CrossRef]
19. Lin, Q.; Painter, O.J.; Agrawal, G.P. Nonlinear optical phenomena in silicon waveguides: Modeling and applications. *Opt. Express* **2007**, *15*, 16604–16644. [CrossRef] [PubMed]
20. Moss, D.J.; Morandotti, R.; Gaeta, A.L.; Lipson, M. New CMOS-compatible platforms based on silicon nitride and Hydex for nonlinear optics. *Nat. Photonics* **2013**, *7*, 597–607. [CrossRef]
21. Blumenthal, D.J.; Heideman, R.; Geuzebroek, D.; Leinse, A.; Roeloffzen, C.G.H. Silicon Nitride in Silicon Photonics. *Proc. IEEE* **2018**, *106*, 2209–2231. [CrossRef]
22. Bristow, A.D.; Rotenberg, N.; Van Driel, H.M. Two-photon absorption and Kerr coefficients of silicon for 850–2200nm. *Appl. Phys. Lett.* **2007**, *90*, 191104. [CrossRef]
23. Soref, R.; Emelett, S.J.; Buchwald, W.R. Silicon waveguided components for the long-wave infrared region. *J. Opt. A Pure Appl. Opt.* **2006**, *8*, 840–848. [CrossRef]
24. Lamy, M.; Finot, C.; Colman, P.; Fatome, J.; Millot, G.; Roelkens, G.; Kuyken, B.; Hammani, K. Silicon waveguides for high-speed optical transmissions and parametric conversion around 2 μm. *IEEE Photonics Tech. Lett.* **2019**, *31*, 165–168. [CrossRef]
25. Liu, X.; Jr, R.M.O.; Vlasov, Y.A.; Green, W.M.J. Mid-infrared optical parametric amplifier using silicon nanophotonic waveguides. *Nat. Photonics* **2010**, *4*, 557–560. [CrossRef]
26. Griffith, A.G.; Lau, R.K.W.; Cardenas, J.; Okawachi, Y.; Mohanty, A.; Fain, R.; Lee, Y.H.D.; Yu, M.; Phare, C.T.; Poitras, C.B.; et al. Silicon-chip mid-infrared frequency comb generation. *Nat. Commun.* **2015**, *6*, 6299. [CrossRef] [PubMed]
27. Pfeiffer, M.H.P.; Herkommer, C.; Liu, J.; Morais, T.; Zervas, M.; Geiselmann, M.; Kippenberg, T.J. Photonic Damascene Process for Low-Loss, High-Confinement Silicon Nitride Waveguides. *IEEE J. Sel. Top. Quantum Electron.* **2018**, *24*, 1–11. [CrossRef]
28. Lamy, M.; Finot, C.; Fatome, J.; Weeber, J.C.; Millot, G.; Kuyken, B.; Roelkens, G.; Brun, M.; Labeye, P.; Nicoletti, S.; et al. High speed optical transmission at 2 μm in subwavelength waveguides made of various materials. In Proceedings of the Integrated Photonics Research, Silicon and Nanophotonics, Zurich, Switzerland, 2–5 July 2018.
29. Lacava, C.; Stankovic, S.; Khokhar, A.Z.; Bucio, T.D.; Gardes, F.Y.; Reed, G.T.; Richardson, D.J.; Petropoulos, P. Si-rich Silicon Nitride for Nonlinear Signal Processing Applications. *Sci. Rep.* **2017**, *7*, 1–13. [CrossRef]

30. Krückel, C.J.; Fülöp, A.; Klintberg, T.; Bengtsson, J.; Andrekson, P.A.; Torres-Company, V. Linear and nonlinear characterization of low-stress high-confinement silicon-rich nitride waveguides: Erratum. *Opt. Express* **2017**, *25*, 7443–7444. [CrossRef]
31. Liu, X.; Pu, M.; Zhou, B.; Krückel, C.J.; Fülöp, A.; Torres-Company, V.; Bache, M. Octave-spanning supercontinuum generation in a silicon-rich nitride waveguide. *Opt. Lett.* **2016**, *41*, 2719. [CrossRef]
32. Ye, Z.; Fülöp, A.; Helgason, Ó.B.; Andrekson, P.A.; Torres-Company, V. Low-loss high-Q silicon-rich silicon nitride microresonators for Kerr nonlinear optics. *Opt. Lett.* **2019**, *44*, 3326–3329. [CrossRef]
33. Mitrovic, M.; Guan, X.; Ji, H.; Oxenløwe, L.K.; Frandsen, L.H. Four-wave mixing in silicon-rich nitride waveguides. In Proceedings of the Frontiers in Optics, San Jose, CA, USA, 18–22 October 2015.
34. Wang, X.; Guan, X.; Gao, S.; Hu, H.; Oxenløwe, L.K.; Frandsen, L.H. Silicon/silicon-rich nitride hybrid-core waveguide for nonlinear optics. *Opt. Express* **2019**, *27*, 23775–23784. [CrossRef]
35. Ooi, K.J.A.; Ng, D.K.T.; Wang, T.; Chee, A.K.L.; Ng, S.K.; Wang, Q.; Ang, L.K.; Agarwal, A.; Kimerling, L.C.; Tan, D.T.H. Pushing the limits of CMOS optical parametric amplifiers with USRN:Si7N3 above the two-photon absorption edge. *Nat. Commun.* **2017**, *8*, 13878. [CrossRef]
36. Wang, T.; Ng, D.K.T.; Ng, S.-K.; Toh, Y.-T.; Chee, A.K.L.; Chen, G.F.R.; Wang, Q.; Tan, D.T.H. Supercontinuum generation in bandgap engineered, back-end CMOS compatible silicon rich nitride waveguides. *Laser Photonics Rev.* **2015**, *9*, 498–506. [CrossRef]
37. Choi, J.W.; Sohn, B.-U.; Chen, G.F.R.; Ng, D.K.T.; Tan, D.T.H. Soliton-effect optical pulse compression in CMOS-compatible ultra-silicon-rich nitride waveguides. *APL Photonics* **2019**, *4*, 110804. [CrossRef]
38. Sahin, E.; Blanco-Redondo, A.; Xing, P.; Ng, D.K.T.; Png, C.E.; Tan, D.T.H.; Eggleton, B.J. Bragg Soliton Compression and Fission on CMOS-Compatible Ultra-Silicon-Rich Nitride. *Laser Photonics Rev.* **2019**, *13*, 1900114. [CrossRef]
39. Lamy, M.; Finot, C.; Parriaux, A.; Lacava, C.; Bucio, T.D.; Gardes, F.Y.; Millot, G.; Petropoulos, P.; Hammani, K. Si-rich Si nitride waveguides for optical transmissions and toward wavelength conversion around 2 µm. *Appl. Opt.* **2019**, *58*, 5165–5169. [CrossRef]
40. Cheng, M.-C.; Chang, C.-P.; Huang, W.-S.; Huang, R.-S. Ultralow-stress silicon-rich nitride films for microstructure fabrication. *Sens. Mater.* **1999**, *11*, 349–358.
41. Jin, Q.; Lu, J.; Li, X.; Yan, Q.; Gao, Q.; Gao, S. Performance evaluation of four-wave mixing in a graphene-covered tapered fiber. *J. Opt.* **2016**, *18*, 075502. [CrossRef]
42. Yap, K.P.; Delage, A.; Lapointe, J.; Lamontagne, B.; Schmid, J.H.; Waldron, P.; Syrett, B.A.; Janz, S. Correlation of Scattering Loss, Sidewall Roughness and Waveguide Width in Silicon-on-Insulator (SOI) Ridge Waveguides. *J. Light. Technol.* **2009**, *27*, 3999–4008. [CrossRef]
43. Payne, F.P.; Lacey, J.P.R. A theoretical analysis of scattering loss from planar optical waveguides. *Opt. Quantum Electron.* **1994**, *26*, 977–986. [CrossRef]
44. Afshar, S.; Monro, T.M. A full vectorial model for pulse propagation in emerging waveguides with subwavelength structures part I: Kerr nonlinearity. *Opt. Express* **2009**, *17*, 2298–2318. [CrossRef]
45. Tan, D.T.H.; Ooi, K.J.A.; Ng, D.K.T. Nonlinear optics on silicon-rich nitride—A high nonlinear figure of merit CMOS platform [Invited]. *Photonics Res.* **2018**, *6*, B50–B66. [CrossRef]
46. Ebendorff-Heidepriem, H.; Petropoulos, P.; Asimakis, S.; Finazzi, V.; Moore, R.C.; Frampton, K.; Koizumi, F.; Richardson, D.J.; Monro, T.M. Bismuth glass holey fibers with high nonlinearity. *Opt. Express* **2004**, *12*, 5082–5087. [CrossRef]

Publisher's Note: MDPI stays neutral with regard to jurisdictional claims in published maps and institutional affiliations.

© 2020 by the authors. Licensee MDPI, Basel, Switzerland. This article is an open access article distributed under the terms and conditions of the Creative Commons Attribution (CC BY) license (http://creativecommons.org/licenses/by/4.0/).

Review

Optical Realization of Wave-Based Analog Computing with Metamaterials

Kaiyang Cheng [1], Yuancheng Fan [2,*], Weixuan Zhang [3], Yubin Gong [1,4], Shen Fei [1,*] and Hongqiang Li [5,*]

1. School of Electrical Engineering and Intelligentization, Dongguan University of Technology, Dongguan 523808, China; chengky@dgut.edu.cn (K.C.); ybgong@uestc.edu.cn (Y.G.)
2. Key Laboratory of Light Field Manipulation and Information Perception, Ministry of Industry and Information Technology and School of Physical Science and Technology, Northwestern Polytechnical University, Xi'an 710129, China
3. Beijing Key Laboratory of Nanophotonics & Ultrafine Optoelectronic Systems, School of Physics, Beijing Institute of Technology, Beijing 100081, China; 2120141404@bit.edu.cn
4. National Key Lab on Vacuum Electronics, University of Electronic Science and Technology of China (UESTC), Chengdu 610054, China
5. Key Laboratory of Advanced Micro-Structure Materials (MOE) and School of Physics Science and Engineering, Tongji University, Shanghai 200092, China
* Correspondence: phyfan@nwpu.edu.cn (Y.F.); shenfei@dgut.edu.cn (S.F.); hqlee@tongji.edu.cn (H.L.)

Abstract: Recently, the study of analog optical computing raised renewed interest due to its natural advantages of parallel, high speed and low energy consumption over conventional digital counterpart, particularly in applications of big data and high-throughput image processing. The emergence of metamaterials or metasurfaces in the last decades offered unprecedented opportunities to arbitrarily manipulate the light waves within subwavelength scale. Metamaterials and metasurfaces with freely controlled optical properties have accelerated the progress of wave-based analog computing and are emerging as a practical, easy-integration platform for optical analog computing. In this review, the recent progress of metamaterial-based spatial analog optical computing is briefly reviewed. We first survey the implementation of classical mathematical operations followed by two fundamental approaches (metasurface approach and Green's function approach). Then, we discuss recent developments based on different physical mechanisms and the classical optical simulating of quantum algorithms are investigated, which may lead to a new way for high-efficiency signal processing by exploiting quantum behaviors. The challenges and future opportunities in the booming research field are discussed.

Keywords: analog optical computing; metamaterials; metasurfaces; quantum algorithm; edge detection

1. Introduction

Exploring novel approaches to improve the computational capacity and efficiency is a goal that humans have been continuously pursuing. Early computers are constructed mechanically [1] or electronically [2–4] on the principle of analog, aiming to perform mathematical operations. Despite the impressive success achieved in the fields of weather prediction, aerospace and nuclear industry, these machines faced significant obstacles of slow response and large size [5]. In the 20th century, digital computing emerged with a rapid development of semiconductor technology and large-scale integrated circuits. It began to gradually substitute for their conventional analogue counterparts on the strength of easy programmability, high speed and flexibility. However, in solving specialized computational tasks, such as imaging-processing and edge detection, digital computers are often inefficient and hindered by high-power consumptions. As Moore's law is approaching its physical limitations, the long-abandoned analog approach as an alternative paradigm has attracted renewed attention for its potential abilities to overcome these shortcomings [6–8].

Analog optical computing offers unique advantages of real-time, power-efficient, high-throughput imaging processing abilities originating from the wave-based nature [7,9–11]. Compared with standard digital processes, all-optical analog approaches do not refer to photoelectric converters but directly manipulate optical signals both temporally and spatially [12]. In the temporal domain, high-speed pulse waveform modulation enables numerous applications, including analog computing [13–19], differential equations solving [19–23], optical memory [24,25], photonic neural networks [26,27] and complex nonlinear system simulation [28–30]. In the spatial domain, the input functions are indicated by the spatial wavefronts which will be mathematically transformed with pre-designed optical systems. Therefore, this computing platform has intrinsic parallel characteristic, showing great potential for accelerating the processing of megalo-capacity datasets and images [31–33].

In addition to classical mathematical operations, researchers have also extended the concept of analog computing to quantum algorithms that promise an exponential speedup that is far beyond classical ones in solving problems of large integer factorization (Shor's algorithm) [34] and combinatorial optimization (Grover's algorithm) [35,36]. Some fundamental properties of quantum computing, including superposition principle and interference phenomena, are the essence of wave nature, which are not exclusive to quantum mechanical but are common to classical waves. By encoding quantum bit (qubits) into different degrees of freedom for the electromagnetic field (e.g., frequency, polarization, orbital angular momentum, space and time bins), many quantum computations can be efficiently simulated in optics [37–68].

However, traditional analog optical computing requires bulky optical components, such as diffractive lenses and spatial filters, which are inconvenient for miniaturization and integration of modern ultracompact optics. Empowered by recent development of nanofabrication technologies, metamaterial [69–79] or its two-dimensional counterpart, metasurface [80–99], is able to tailor subwavelength building blocks on the scales of micro- or nanometers, providing unprecedented flexibility to arbitrarily control the electromagnetic waves that are unattainable in the nature. Owning to their powerful wave manipulation abilities and deeply subwavelength characteristics, those meta-structures can significantly reduce the complexity and shrink the size of computing systems, making it possible to realize chip-level all-optical information processing systems [100,101].

In this review, we will focus on recent developments in both the physics and applications of the spatial analog optical computing. The paper is organized in four parts as described below. In Section 2 we overview the basic concept of computational metamaterials and the general principles for design and implementation for classical mathematical operations including integration, differentiation and integration equation solution. In Section 3 we introduce a type of wave-based signal processors that can mimic quantum mechanism with classical optical waves. These studies show that the quantum algorithms like Grover's search algorithm and Deutsch-Jozsa algorithm can be simulated by cascading metamaterial functional blocks. In the last section, we conclude with an overview of computational metamaterials based on different mechanisms, the main challenges and the opportunities in the field for future research.

2. Computational Metamaterials

In 2014, Silva et al. [102] proposed a concept of "computational metamaterials" by locally tailoring the electromagnetic parameters of the metamaterials, the elaborately designed meta-structures can reshape the spatial profile of the input signal to perform mathematical operations including spatial differentiation, integration and convolution. The basic idea of computational metamaterials is schematically illustrated in Figure 1a as an example. The stacked multi-layered structure is functional as a first-order differentiator, for arbitrary input wavefronts $f_1(y)$ and $f_2(y)$, the corresponding output profiles are proportional to $df_1(y)/dy$ and $df_2(y)/dy$, respectively. Compared with conventional analog signal processors and Fourier optics system, metamaterial-based approach provides more flexible mechanism for manipulation, more integrable volume and subwavelength thickness.

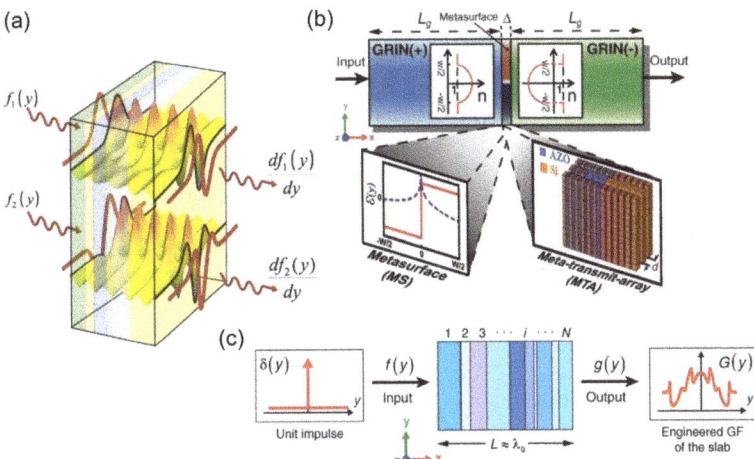

Figure 1. (a) The conceptual sketch of computational metamaterials; schematic of the general protocol for (b) MS approach; and (c) GF approach; adapted with permission from [102], AAAS, 2014.

In general, there are two major protocols to realize metamaterial-based analog computing: metasurface (MS) approach and Green's function (GF) approach. Similar to the classical 4f systems, the MS approach is based on spatial Fourier transformation, with a single-layered metasurface or multi-layered meta-transmit(reflect)-array to realize the desired transfer function, sandwiched between two subblocks performing the Fourier and inverse Fourier transform (see Figure 1b). In the GF approach, by optimizing the transmitted (reflected) response of the metamaterial slabs it can also act as certain mathematical operations without involving additional Fourier lenses. The former has the advantage of implementation simplicity and the latter has a more compact size. In the following sections, we will briefly review the design principle and recent progress pioneered by these two approaches.

2.1. Metasurface Approach

Initially, let us consider a pure mathematical transformation of convolution operation in Fourier domain. For an given input function $f(x,y)$, the corresponding output function $g(x,y)$ is the result of a desired mathematical operator which is indicated by the Green's function $h(x,y)$, x and y denote the spatial coordinates on the two-dimensional (2D) plane. By applying the linear convolution operation, $g(x,y)$ can be described by

$$g(x,y) = \iint f(x',y') h(x-x', y-y') dx' dy' \tag{1}$$

Based on convolution theorem, the Fourier transform of $g(x,y)$ is equivalent to the multiplication of functions f and h in the Fourier domain:

$$g(x,y) = \mathcal{F}^{-1}\{H(k_x,k_y)\mathcal{F}[f(x,y)]\} \tag{2}$$

where $(\mathcal{F}^{-1}\{\cdot\})\mathcal{F}\{\cdot\}$ represents the (inverse) Fourier transform, abbreviated (I)FT. k_x and k_y are frequency variables in the spatial Fourier space, $H(k_x,k_y)$ is spatial FT of the transfer function $h(x,y)$. It is actually easy to implement Equation (2) in the practical systems by simulating the input (output) function with a classical electric field $E_{in}(x,y)$ ($E_{out}(x,y)$), in this case, the spatial variables (x,y) can play the role of (k_x,k_y). In wave-based computing system, Equation (2) can be rewritten as:

$$E_{out}(x,y) = \mathcal{F}^{-1}\{H(x,y)\mathcal{F}[E_{in}(x,y)]\} \tag{3}$$

According to Equation (3), the optical computing system should comprise of three cascade metamaterial functional subblocks which are designed to accomplish FT, $H(x,y)$ and IFT sequentially. In order to realize the FT subblock, an analog of traditional lenses is employed since the converging lenses can achieve 2D Fourier transform in the focal plane. In [102], a graded-index (GRIN) dielectric slab with a parabolic variation of permittivity is introduced to realize FT operator [103,104]. Besides of GRIN medium, meta-lens can also be applied to perform FT operator [105–110]. It worth nothing that, although we can define a GRIN(−) subblock to perform IFT in which the effective constitutive parameters should satisfy $\varepsilon(\mu)_{GRIN(-)}(x,y) = -\varepsilon(\mu)_{GRIN(+)}(x,y)$, it is not feasible for natural materials or practical for metamaterials to realize simultaneously negative permeability and permittivity. Therefore, as an alternative method, we can use additional FT subblock to approximate output profile according to the relation:

$$\mathcal{F}\{\mathcal{F}\{E_{out}(x,y)\}\} \propto E_{out}(-x,-y) \tag{4}$$

In this way, the electric field distribution is proportionate to the mirror image of the desired output function. Based on above discussion, the fundamental components of MS approach are represented in Figure 1b. Next, we will go to the details of how to design transfer functions when implementing different mathematical operations, such as differentiation, integration and convolution.

Firstly, for simplicity, we discuss the cases of one-dimensional nth derivation operators. Based on derivative property of the Fourier transform, $d^n f(y)/dy^n = \mathcal{F}^{-1}\{(ik_y)^n \mathcal{F}[f(y)]\}$, according to Equation (3), the expression $(ik_y)^n$ is the required transfer function, where $H(y) \propto (ik_y)^n$ or $H(y) \propto (iy)^n$. It means that the amplitude and phase of impinging optical wave will be modulated by $e^{ik\Delta} \propto (iy)^n$ when propagating through the transfer function subblock (suppose the thickness of transfer function subblock is Δ), where $k = (2\pi/\lambda_0)\sqrt{\varepsilon_\Delta(y)\mu_\Delta(y)}$, with free space wavelength λ_0, ε_Δ and μ_Δ are relative permittivity and permeability of transfer function subblock. For the cases of integration and convolution operations, the transfer functions become to $H(y) \propto (iy)^{-1}$ and $H(y) \propto \sin c(y)$, respectively. In addition, considering the limitation of lateral dimension, the transfer function $H(y)$ has to be normalized to ensure the energy conservation (without gain materials) and keep the reflection (transmission) coefficient blow unity. Now, the question is, how to find a proper spatially-variant meta-structure that fulfills the required electromagnetic response. Here, we divide recent achievements of the MS approach into two categories according to the construction type.

2.1.1. Reflective Configurations of MS Approach

Plasmonic metasurfaces [83,111–119] promise the abilities to tailor the local phase and amplitude by exploiting the strong light-matter interactions due to the plasmonic resonances. In 2015, an inhomogeneous plasmonic metasurface approach, which consists of three-layer sandwich structure was introduced to optical analog computing for the first time by Anders Pors et al. [120]. A periodic arrangement of gold nanobrick arrays is on the top layer, an optically thick gold ground and subwavelength dielectric spacer is used to excite gap-surface plasmon (see Figure 2a). By properly designing the position and size of each nanobrick, this metal-insulator-metal structure can perform integration at visible wavelength with high efficiency. Similar to this work, in 2017, Chen et al. [121] used a dendritic structure instead of nanobrick to implement first-order differential operation. By varying the geometrical shapes, the desired reflection coefficients could be achieved (see Figure 2c). A different approach for reflectarray configuration is dielectric metasurface, which is composed of high refractive index dielectric nanoparticles supporting both magnetic and electric dipole-like resonances based on Mie theory. This approach can significantly reduce the ohmic losses caused by metallic structures in the optical spectrum. Motivate by this, Ata Chizari et al. [122] propose a dielectric meta-reflect-array where silicon nanobricks are placed on the silica spacer and silver substrate to manipulate amplitude and phase of reflected cross-polarized field by varying the lengths of nanobricks (see Figure 2b).

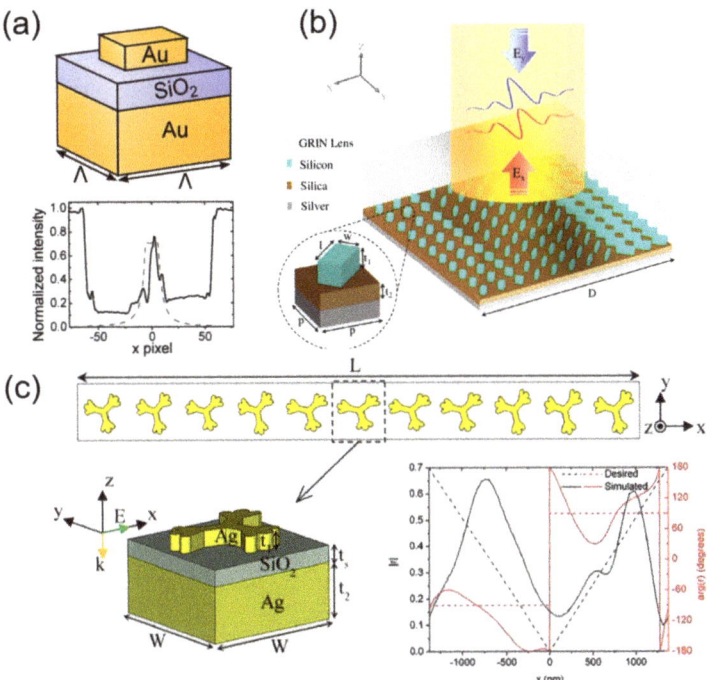

Figure 2. (**a**) The unit cell of plasmonic metasurfaces consist of gold nanobrick atop a dielectric spacer and thick metal ground; adapted with permission from [120], ACS Publications, 2015. (**b**) Schematic representation of dielectric meta-reflect-array for the first-order derivation; adapted with permission from [122] © The Optical Society. (**c**) Top: eleven silver dendritic structure units with different parameters; left: the unit cell of plasmonic metasurfaces with dendritic structure instead of nanobrick; right: comparison of the theoretical and simulated results of the position-dependent reflection coefficient. Adapted with permission from [121] © The Optical Society.

2.1.2. Transmittive Configurations of MS Approach

Compared with reflective configurations, the optical computing systems with transmission mode are easily applied to optical devices, however most of them are facing the challenges of low transmission efficiency. In 2015, Amin Khavasi et al. [32] proposed a concept of "metalines" in which three symmetrically stacked graphene-based building blocks were utilized to perform differentiation and integration in the transmitted way (see Figure 3a). By manipulating plasmons surface wave via surface conductivity variation of the graphene, the amplitudes and phases of the transmitted wave could be controlled completely. While it was a proof-of-principle study, it opened up an avenue of the graphene-based structure which are designed to implement the mathematical operators with advantages of dynamic tunable and high-compact characteristics. Then, in 2016, Zhang et al. [123] introduced a more practical and flexible computing metamaterial system based on effective medium theory by drilling sub-wavelength hole-arrays with different radiuses, the desired effective permittivity distribution can be obtained to realized functional subblocks (see Figure 3b). To further improve the parallel processing abilities, a multi-way analog computing device combined with additional transformation optical subblock is investigated [124]. Two identical signals on the same input port propagate along opposite directions and perform first- and second-order differentiators simultaneously. Very recently, a one-dimensional high-contrast transmitarray (HCTA) metasurface was experimentally demonstrated by Zhang et al. [100] to realize Fourier transform and spatial differentiation with feature size of 140 nm. The on-chip meta-system provides low insertion loss and broad operating bands by tailoring widths and lengths of the rectangular slot arrays (see Figure 3c,d). Inspired by radial Hilbert transform filter, Huo et al. [125] proposed a spin-dependent dielectric metasurface, which can perform a spiral phase filtering operation in which a spiral phase profile is imprinted to the input wave. The tricky part in this proposal is to utilize the π phase difference in opposite azimuth leading to a destructive interference when processing convolution transformation. The authors demonstrate that spiral phase filtering operation is equivalent to the 2D spatial differentiation of incident light field (see Figure 3e).

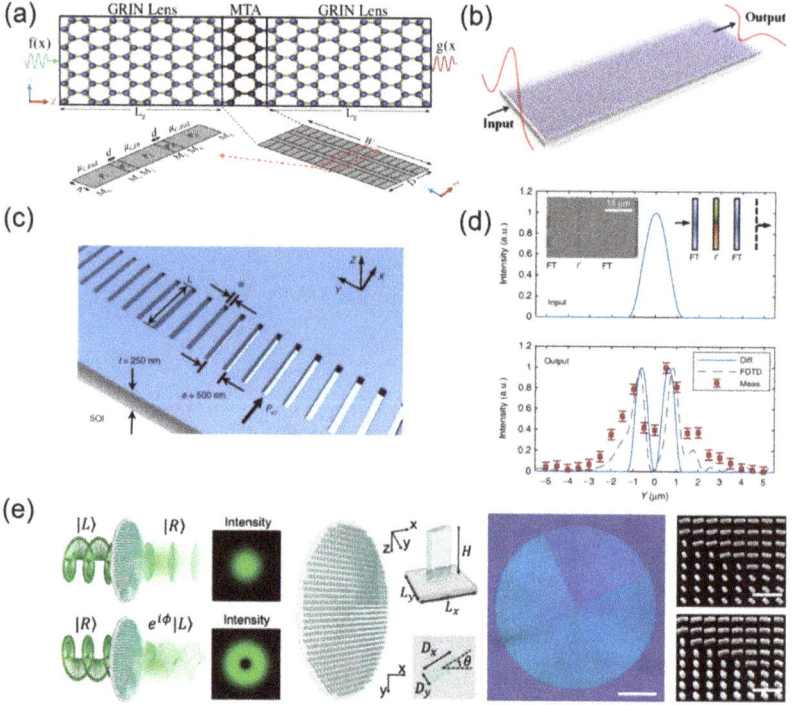

Figure 3. (a) The sketch of graphene-based metaline; adapted with permission from [32] © The Optical Society. (b) Dielectric metamaterial with drilling hole arrays to solving second-order differential equation; adapted with permission from [123], IOP Publishing, 2016. (c) Schematic of on-chip 1D high-contrast transmitarray, the rectangular slot arrays are etched on the silicon-on-insulator (SOI) substrate with the period of 500 nm. (d) Comparison of the input (top) and output (bottom) intensity profile of an on-chip differentiator; adapted with permission from [100], Springer Nature, 2019. (e) Left: schematic of the concept for spin-dependent function control; middle: schematic of the dielectric metasurface spatial filter; right: photograph of the fabricated metasurface. Scale bar: 500 µm. Insets: scanning electron micrographs showing the top and oblique view of TiO_2 nanopillar array. Scale bar: 1 µm. Adapted with permission from [125], ACS Publications, 2020.

2.2. Green's Function Approach

In the original GF approach proposed by Silva et al. [102], the transmission or reflection coefficient of multi-layered stacked slabs with transversely homogenous and longitudinally inhomogeneous are engineered to approximate the desired transfer function without getting into the Fourier domain. Since additional Fourier lenses are not required, the GF approach has advantages over the MS approach with a more compact size and an easier fabrication process. The basic idea of the GF approach is to find the proper transmittance coefficient $\widetilde{T}(k_y)$ of the multi-layered slab to fit the desired GF kernel $\widetilde{G}(k_y)$ within $0 < k_y < k_0$, in which k_y is the transverse wavevector, and k_0 is the free space wavevector. A fast synthesis algorithm is adopted to minimize the difference between $\widetilde{T}(k_y)$ and $\widetilde{G}(k_y)$, which can be expressed as:

$$err = \sum_{i=1}^{M} \left(w_r \left(\text{Re}\left[\widetilde{G}(k_{y,i}) - \widetilde{T}(k_{y,i}) \right] \right)^2 + w_i \left(\text{Im}\left[\widetilde{G}(k_{y,i}) - \widetilde{T}(k_{y,i}) \right] \right)^2 \right) \quad (5)$$

where M is the number of layers, w_r and w_i are weight coefficients for real and imaginary parts. The optimized transmission spectrum is in good agreement with the desired GF which correspond to 2th spatial derivative provided that M = 10, w_r = 2, w_i = 1, and relative permeability of each layer keep at unity.

In recent years, extensive studies with resonant or non-resonant structures were proposed with the GF approach, such as multi-layered slabs [126,127], photonic crystal [128], diffraction gratings [15,31,129–134], surface plasmon-based devices [130,135–137], dielectric metasurfaces [138–143], nonlocal metasurfaces [144,145], random medium [146] and inverse-designed meta-structures [147]. Moreover, various physical mechanisms including Brewster effect [148], Goos-Hänchen effect [149] and spin hall effect [150] can also be applied to perform optical analog computing. Therefore, the GF approach can be generalized to a fundamental protocol for designing the particular transfer function (defined as $H(k_x, k_y) = \widetilde{E}_y^{out}(k_x, k_y) / \widetilde{E}_y^{in}(k_x, k_y)$) which should be approximately proportional to the desired mathematical operations.

2.2.1. Diffraction Gratings

Diffraction gratings have been widely used in processing optical signals with integration or differentiation operations, most of them are based on Fano resonance or guided-mode resonance in which the reflection or transmission coefficient can approximate the transfer function in the vicinity of the resonance. In [131], Victor A. Soifer et al. experimentally demonstrate that a guided-mode resonant grating with period TiO_2 slab waveguide on a quartz substrate can perform spatial differentiation with an obliquely incident beam. As shown in Figure 4a, the transverse profile of the incident wave $P_{inc}(x, y)$ is formed by zeroth order diffraction to the transmitted profile of $p_{tr}(x, y)$. Figure 4b is the scanning electron microscopy image of the designed diffraction grating. A similar strategy is to adopt high-contrast grating, as illustrated in Figure 4c [132]. By observing the amplitude and phase (indicated by red dot and blue line in Figure 4d) of the spatial spectral transfer function evolving with different incident angle, it is found that the transfer function has a phase variation of π around a certain angle θ_0. Within slight deviation of θ_0, the transfer function satisfies first-order differentiation. In a recent work, Alexios Parthenopoulos et al. [134] propose a novel dielectric subwavelength grating by employing high quality suspended Si_3N_4 films, where the first- and second-order spatial differentiation can be implemented at oblique and normal incidence, respectively. The even guided mode is excited with normal incident wavefront, and change dramatically to the odd guided mode as increasing the incidence angle. In addition to these dielectric gratings, metallic gratings are also investigated and experimentally demonstrated to realize analogue spatial differentiators. It is well-known that when a metallic grating is illuminated by an incident wave, at certain incident angle, the surface plasmon resonance will be excited. Yang et al. [133] found that by adjusting the geometric parameters of the subwavelength gratings, in the vicinity of surface plasmon resonance, the transfer function can be tailored to a linear function which satisfies the requirements of first-order differentiation.

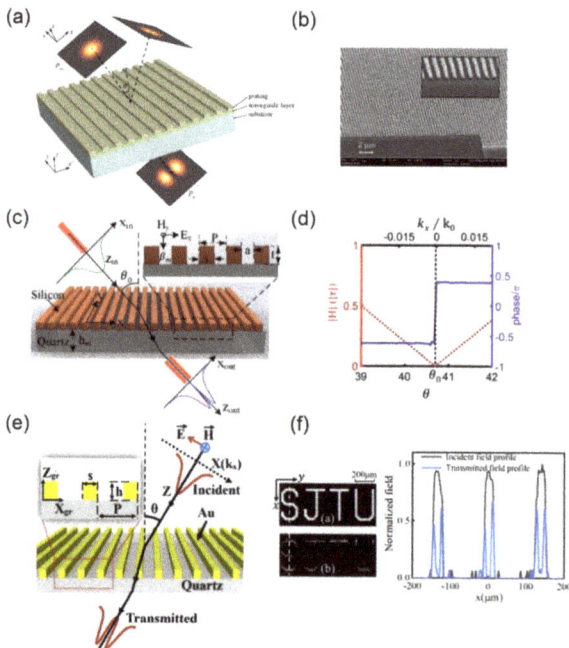

Figure 4. (a) Schematic and (b) photograph of guided-mode resonant diffraction grating for first-order differentiation; adapted with permission from [131] © The Optical Society. (c) The schematic and (d) spatial spectral transfer function of the optical spatial differentiator based on subwavelength high-contrast gratings. Adapted with permission from [132], AIP Publishing, 2018. (e) Schematic of the first-order differentiator based on subwavelength metallic gratings. (f) The experimental results for the incident and transmitted images (left) and the normalized field profiles at the position of white dashed line. Adapted with permission from [133] © The Optical Society.

2.2.2. Plasmonic Structure

By exploiting the critical coupling condition of the surface plasmon polaritons (SPPs), many plasmonic analog computing devices have been realized. In 2017, Zhu et al. [135] introduced a classical Kretschmann prism configuration to realize the plasmonic spatial differentiation where the oblique incident wave (transverse profile of $S_{in}(k_x)$) with TM-polarized are used to excite the surface plasmon between glass substrate and silver film, k_x is the transverse component of the wavevector (see Figure 5a). In this case, the amplitude of reflected wave (transverse profile of $S_{out}(k_x)$) is the result of the interference between the direct reflection and the leakage of SPPs. The spatial spectral transfer function around $k_x = 0$ can be expressed as:

$$H(k_x) = e^{i\varphi}\frac{ik_x + A}{ik_x + B} \tag{6}$$

where φ is the phase change related to the process of the direct reflection at the glass–metal interface. $A = (\alpha_{spp} - \alpha_1)/\cos\theta_0$ and $B = (\alpha_{spp} + \alpha_1)/\cos\theta_0$, where α_1 and α_{spp} are the radiative leakage rate of the SPP and the intrinsic material loss rate, respectively. When the critical coupling condition is satisfied ($\alpha_1 = \alpha_{spp}$), Equation (6) can be simplified and approximated as $H(k_x) \propto (e^{i\varphi}/B)ik_x$, which corresponds to the first-order differentiation.

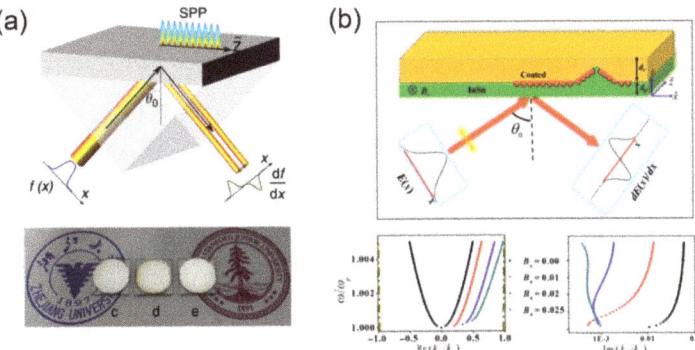

Figure 5. (a) Top: schematic of the surface-plasmon-based scheme to perform spatial differentiation with the Kretschmann configuration. Bottom: photograph of the three samples with different thicknesses; adapted with permission from [135], Springer Nature, 2017. (b) Top: schematic of the backscattering-immune first-order spatial differentiator based on nonreciprocal plasmonic platform. Bottom: the simulated results of real and imaginary parts of the eigenvector with different external magnetic fields. Adapted with permission from [137], APS, 2019.

However, the SPPs is sensitive to the defect of the interface. As such, to further improve its robustness, Zhang et al. [137] proposed a unidirectional SPPs-based spatial differentiator which can avoid backscattering by applying an external static magnetic field (shown in Figure 5b). The magnetic field breaks the time-reversal symmetry and makes the system have a nonreciprocal property, which is critical to the unidirectional SPPs leaky mode. As shown in the bottom of Figure 5b, when applying external magnetic field, the SPPs surface mode changes from bidirectional to unidirectional. To verify the practicability when facing real-time image processing, the impact of plasmonic spatial differentiation under time-variant optical signals is analyzed by Zhang et al. [136]. By adopting similar setup described in [135], the reflected field profile with central frequency ω_0 in the spatiotemporal coordinate can be expressed as:

$$s_{out}(x,t) = c_s \frac{ds_{in}(x,t)}{dx} + c_t \frac{ds_{in}(x,t)}{dt} \tag{7}$$

where c_s and c_t are two constants where $c_s = e^{i\phi}\frac{\cos\theta_0}{2a_1}$ and $c_t = e^{i\phi}(2a_1)^{-1}\left(\frac{1}{v_g} - \frac{\sin\theta_0}{v_{gls}}\right)$, ϕ is phase change of direct reflection without excitation of SPP, θ_0, a_1, v_g, v_{gls} are the incident angle, the leakage rate, the group velocity of the SPP, and the speed of light in the prism at central frequency ω_0, respectively. The second term of Equation (7) indicates that the output field profile should consider the change rate of input time-modulated signals. As a conclusion, the authors estimate that the processing speed of plasmonic differentiator is up to the maximum of 10^{13} frame/s which is restricted by the time of establishing the SPP leaky radiation.

2.2.3. Two-Dimensional Dielectric Metasurfaces

The above discussions are mainly focused on 1D mathematical operations. However, anisotropic edge detection is not sufficient for real-time, high-throughput 2D image processing since multiple measurements are required. In this part, metasurfaces with symmetrically distributed dielectric nano-resonators are briefly reviewed to illustrate their capacity to reveal all information from the boundaries of objects. The basic physical mechanism of these 2D dielectric metasurfaces originates from the modal evolution where a bound state resonance mode with normal incident plane wave will change to a leaky waveguide mode at oblique incidence. The transmitted amplitudes encoded with information of spatial variation are sensitive to the transmission when increasing the incidence angle [151].

A type configuration is illustrated in Figure 6a [143], silicon nitride patch-arrays are embedded in a homogeneous medium of silicon dioxide. By tuning the height of patch-arrays, the spatial bandwidth and resolution of the edge detection can be modulated. The transmission coefficient of TE incident wave can be expressed as $T_{TE}(\varphi, k_r) \approx \alpha(\varphi) k_r^2$, where φ is the azimuthal angle, $\alpha(\varphi)$ is the gain of the system, and $k_r = \sqrt{k_x^2 + k_y^2}$. From Figure 6b, it can be seen that the transmission coefficient on the xoy plane formed the symmetric parabolic distribution due to the geometric symmetric, thereby it can be used to perform the second-order derivative with the quadratic approximation (see Figure 6c). In [142], a similar strategy is adopted to realize a 2D Laplace operator which can be combined with traditional imaging systems with a numerical aperture (NA) up to 0.32. However, since the quasi-guided leaky modes can only couple with p-polarized incident waves due to the modal symmetry, this scheme can only operate for one polarization. To overcome this limitation, Wan et al. [141] introduce a spatial differentiator can be performed for arbitrary polarized (unpolarized) wave by tailoring the spatial dispersion of electric dipole resonance supported by silicon nanodisks (see Figure 6d). Figure 6e shows that for both s- or p-polarized incident waves, the transfer function have similar parabolic shape which indicate the functionality of spatial differentiation for arbitrary polarization. The corresponding experiments verify the performance of this scheme (see Figure 6f).

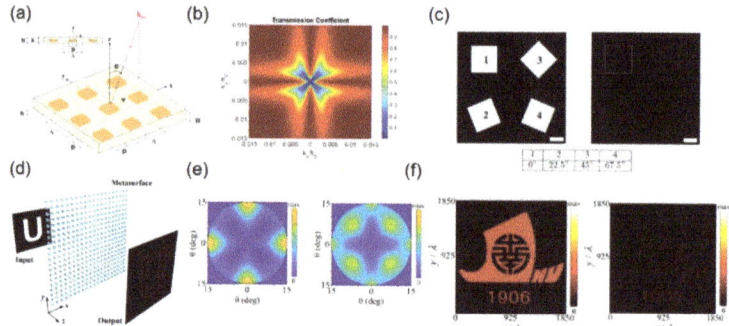

Figure 6. (a) Schematic of the square patch-arrays metasurface for 2D edge detection; (b) 2D spatial transfer function spectrum; (c) edge detection of square-shaped input beams; adapted with permission from [143], IEEE, 2018. (d) Schematic of the basic configuration of dielectric metasurface for 2D spatial differentiation; (e) the transfer functions for or both linearly polarized light fields; (f) the output image of 2D edge detection for x polarization light waves. Adapted with permission from [141] © The Optical Society.

3. Other Emerging Approaches
3.1. Nonlocal Metasurface

Kwon et al. [144] introduce a new mechanism with increasing the nonlocal response of metasurfaces, which is generally considered to be undesirable and detrimental, by sinusoidally modulating the permittivity of each split-ring resonator (SRR) (Figure 7a). Within the modulation, the transmission formed a Fano response where a sharp variation of transmission coefficient emerges on the resonance frequency accompanied with the incident angle (see Figure 7b). By adding a horizontally misplaced metallic wire-arrays, the requirements of breaking both vertical and horizontal mirror symmetry are fulfilled to perform first-order derivative operation (see Figure 7c). Furthermore, 2D edge detection is demonstrated with combining the 1D computing metasurface and its 90° rotational symmetric structure (see Figure 7d–f).

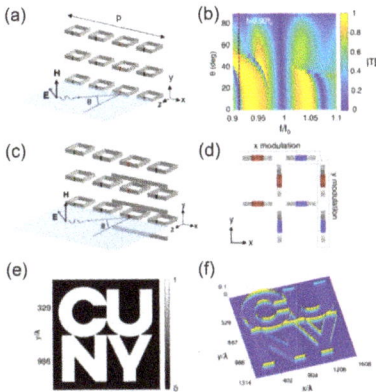

Figure 7. (a) Schematic of sinusoidally modulated of split-ring resonators; (b) transmission spectrum with different incident angle; (c) schematic of asymmetric metasurface with a horizontally misplaced array of metallic wires for the first-derivative operation; (d) rotated metasurface for 2D second-derivative operation. (e) Input image and (f) output image of 2D edge detection for linear polarized wave. Adapted with permission from [144], APS, 2018.

3.2. Random Medium

Another counterintuitive approach is by employing random medium rather than deliberately designed structures with certain scattering properties to perform wave-based analog computing. In the study of Hougne et al. [146], it consists of two subsystems: a chaotic cavity as the random medium and a reflect-array metasurface playing the role of wavefront shaping device. As shown in Figure 8a, a plane impinging wave with input vector X is modulated by a wave front shaping device and subsequently propagating through a random medium to obtain the desired output wave front. For controlling spatial degree of freedom, the incident wave front is divided into four segments, marked by A to D (see Figure 8b). Here, an indoor cavity of irregular geometry is used as the random environment characterized with the Green's function, which plays the key role of analog computing by enabling each segment of wave front to contribute to each output point. The reconfigurable metasurface contains 88 units whose reflection coefficient can be tuned from −1 to 1. Figure 8c shows the experimental configuration in which microwave absorbers are installed to limit the reverberation of the cavity.

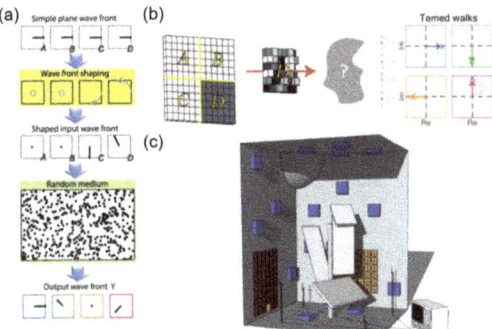

Figure 8. (a) Schematic the random medium scheme for wave-based analog computation; (b) tamed contribution of segment D to observation points; (c) experimental setup in a metallic cavity of irregular geometry. Adapted with permission from [146], APS, 2018.

3.3. Inverse Design

Estakhri et al. [147] implement the discrete numerical method of solving integral equation in an optical way by using a recursive approach with a metamaterial block, feedback waveguides and coupling elements (see Figure 9a). The Fredholm integral equation of the second kind is expressed as: $g(u) = I_{in}(u) + \int_a^b K(u,v)g(v)dv$, where $g(u)$ is the function to be solved, $K(u,v)$ and $I_{in}(u)$ are integral kernel and input signal. By sampling the complex values of electric fields on the input and output plane (with N points), the corresponding $N \times N$ matrix equation can be established. To obtain the function $g(u)$ is equivalent to calculate the inverse $N \times N$ matrix (A), which can be realized by an inhomogeneous permittivity distributed metamaterial block. In the study, the authors adopt the objective-first optimization technique to approach the desired relative permittivity in each iteration constrained by criterion of transfer function I-A, where I is the identity matrix (the simulation results is shown in Figure 9b). The experiment of reflective configuration verifies that it is feasible to apply the inverse-designed metamaterial platform in solve integral equation (see Figure 9c,d).

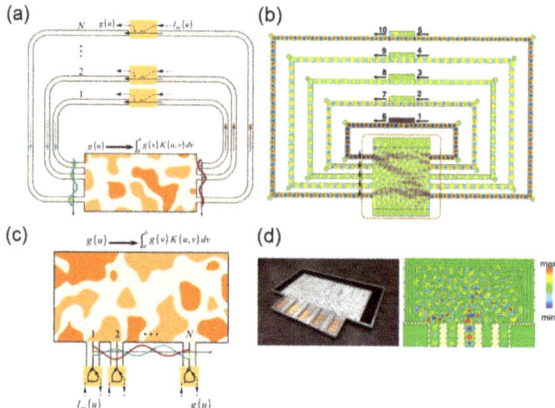

Figure 9. (**a**) The conceptual sketch of the inverse-designed metamaterial platform in solve integral equation; (**b**) time snapshot of the simulation results for the transmissive configuration with external feedback network; (**c**) the reflective configuration of the internal feedback metamaterial system. (**d**) Left: the photograph of the constructed metamaterial system. Right: time snapshot of the simulation results for the reflective configuration with internal feedback network. Adapted with permission from [147], AAAS, 2019.

3.4. Brewster Effect

As mentioned above, the transversely homogenous multilayer slabs with even symmetry in the spatial Fourier domain cannot be applied to perform first-order derivative or integration operators since the odd symmetry is needed for these operations. In the study of Youssefi et al. [148], a rotated configuration is theoretically demonstrated to overcome this constraint by breaking the refection symmetry. The schematic of the rotated structure is shown in Figure 10a, when the oblique incident wave illuminating on the boundary of two dielectric medium (in this study, they are air and a dielectric substrate with different refractive indices of 1 and 2.1) at the Brewster angle, the GF around $k_y = 0$ can be approximated with a linear function which is used to implement first-order derivative (see Figure 10b). The proposed simple structure enables effective implementation of the mathematical operations, however, there are some drawbacks in this configuration. The approximation requires the reflection spectrum of the interface to become equal to zero, which means only the signals around the Brewster angle can perform the derivation, leading to a relatively narrow spatial spectrum and small reflected energy.

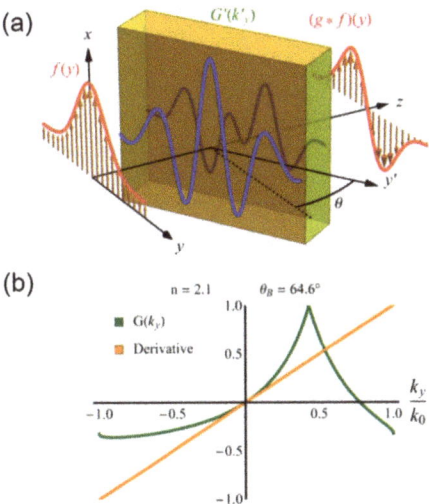

Figure 10. (**a**) The schematic of the rotated structure for realizing an even or odd Green's function; (**b**) the exact GF distribution and its approximation around $k_y = 0$ based on the Taylor series. Adapted with permission from [148] © The Optical Society.

3.5. Goos-Hänchen Effect

Motivated by the high sensitivity of Goos-Hänchen (GH) effect for the interface of total internal reflection, Xu et al. [149] propose a GH-based approach where the kernel transfer function is acting on the air–glass interface to perform first-order derivative. After the total internal reflection, the reflected angular spectrum contains two linearly polarized components which are converted by a quarter-wave plate into two opposite circular-polarized components. The destructive interference of the reflected waves leads to the functionality of edge detection. The experimental setup is illustrated in Figure 11a, the phase distribution and spatial spectral transfer function are shown in Figure 11b. the results of edge detection for different images are shown in Figure 11c.

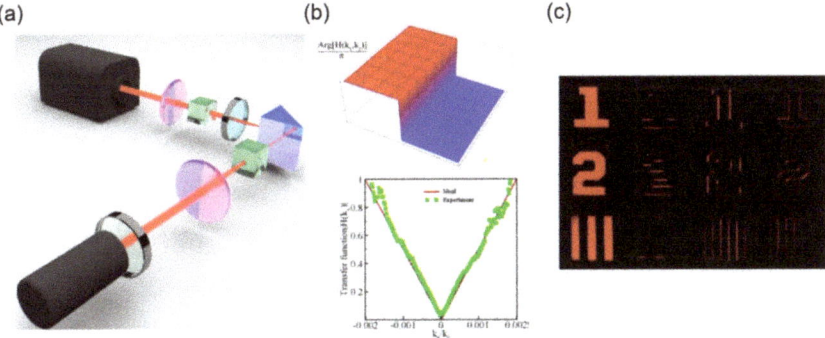

Figure 11. (**a**) The experimental setup for measurement of the spatial spectral transfer function of the first order spatial differentiator based on GH effect; (**b**) top: the phase distribution for the spatial spectral transfer function. Bottom: the comparison of theoretical and experimental results for the spatial spectral transfer function for $k_x = 0$; (**c**) output image of edge detection with different rotating angles. Adapted with permission from [149], AIP Publishing, 2020.

3.6. Spin Hall Effect

In the study of Zhu et al. [150], the authors demonstrate that realizing spatial differentiation is a natural effect of spin Hall effect (SHE) for the reflected or refracted wave at any planar interface. In this proposal, spin-dependent transverse shifts of $-\delta$ and $+\delta$ corresponding to parallel and antiparallel spin states, the destructive interference of the opposite shifts by selecting an orthogonally polarized reflected wave lead to the final output wave function, which can be expressed as: $|\varphi_{out}\rangle = \frac{i}{2}\int dy[\varphi_{in}(y+\delta) - \varphi_{in}(y-\delta)]|y\rangle$, where $|\varphi_{in}\rangle$ and $|\varphi_{out}\rangle$ denote the input and output wave functions. The limit of $|\varphi_{out}\rangle$ is approximately proportional to the first-order spatial differentiation when δ is much smaller than the initial wavefunction profile. As an example, Figure 12a shows that an obliquely incident paraxial beam illuminates on the interface of two isotropic media, two orthogonal polarizers are installed between the incident and reflected waves. The transfer function around $k_y = 0$ is depicted in Figure 12b. the experimental results for 1D edge detection are shown in Figure 12c.

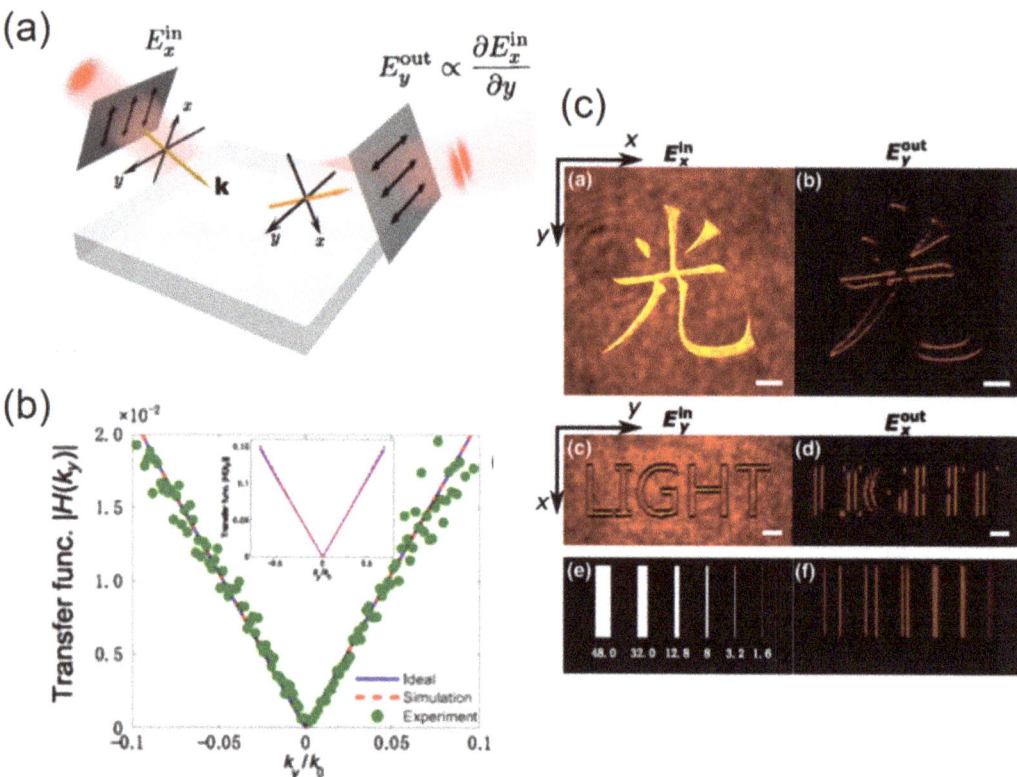

Figure 12. (a) Schematic of spatial differentiation from the SHE of light on an optical planar interface between two isotropic materials; (b) the comparison of theoretical and experimental results for the spatial spectral transfer function for $k_x = 0$; (c) the results of edge detection with different target images stored in $E_{in}x$ and $E_{in}y$, respectively. Adapted with permission from [150], APS, 2019.

3.7. Quantum Computing with Metamaterials

Quantum computation [38], by employing the principles of quantum mechanics, such as superposition and entanglement, provides the basis for quantum algorithms that enable tremendous improvements over classical computing technology for certain complex tasks. For example, the large integer factorization problems are considered practically impossible for classical algorithms as the numbers get larger than 2048 bits. Therefore, it is essential for modern RSA encryption because the codes are virtually unbreakable. However, this cryptosystem is facing a daunting challenge by Shor's algorithm [34], which promises the abilities of factoring integers in polynomial time. Another famous example is Grover's search algorithm [35,36] in which searching an unsorted database only takes $O(\sqrt{N})$ time compared with classical algorithm of $O(N)$, where N is the number of entries in the database. Although the quadratic speedup is slightly less impressive than other quantum algorithms with exponential acceleration capacities, the time saved is significant when N is considerably large. In fact, the extraordinary power of quantum computation comes from the smallest information carrier named quantum bits (or "qubits"). The quantum superposition phenomenon allows a single qubit to hold 0 and 1 states at the same time, as multiple qubits is coherently controlled, they process inherent parallel computing abilities that are far beyond the fastest classical computers.

However, despite these advantages over classical computers, up to now, to realize large-scale universal quantum computer is extreme difficult. Until 2016, a proof-of-principle demonstration of Shor's algorithm can only be achieved to factor the number 15 [152]. Apparently, quantum computation still has a long way to go before it becomes practical. The biggest challenges that we faced were preparing, control and measurement of qubits, actions on qubits should be very carefully handled that even the slightest interaction with surrounding environments may leading to decoherence and destroy the quantum information. Among enormous numbers of experimental realizations of quantum algorithms in different physical systems (including nuclear magnetic resonance, quantum dots, trapped ions or QED cavities), the optical implementation is a promising way since the photons are more robust to external perturbation and thus corresponds to long decoherence times. In addition, the quantum optical system do not need extremely cold temperatures to function. The central idea of quantum optical computing is that quantum information can be encoded by different degrees of freedom of photons (e.g., polarization, orbital angular momentum, spatial and temporal modes), by utilizing the common properties shared by both classical optics and quantum mechanics, such as superposition and interference, it is possible to simulate certain quantum behaviors with classical light [37,39,40,42,43,45,48–53,55–59,62,67,153].

As a new platform for arbitrarily manipulating wavefront of light, metamaterials or metasurfaces can also be applied in simulating quantum algorithms with classical wave-based optics. In 2017, Zhang et al. [154] proposed a metamaterial-based quantum algorithm analog to perform Grover's search algorithm. The designed metamaterial consisted of four cascade subblocks (an oracle subblock U_m, two Fourier transform subblocks F, a phase plate subblock $I - 2|0\rangle\langle 0|$) corresponding to the operators of oracle, Walsh–Hadamard transform and inversion-about-average (IAA) operation in Grover's search algorithm (see Figure 13a). In this scheme, the quantum states are directly mapped onto classical optical fields. For instance, the incident electric field amplitude "$E(y)$" is the analogue of quantum probability amplitude, transversal coordinate "y" is used to label the item of the database and the maximum number of the database is depending on the full width at half-maximum of the beam "D" (listed in Table 1). When light occurs on the metamaterial, each functional subblock it passed is equivalent to performing a quantum operation; therefore, each roundtrip represents one iteration of the search algorithm. After multiple iterations in the metamaterial, the marked item is found by measuring the field distribution on the output plane (see Figure 13b). Recently, Cheng et al. [155] verified that Deutsch-Jozsa (DJ) algorithm could be simulated by using a similar strategy (see Figure 13c,d).

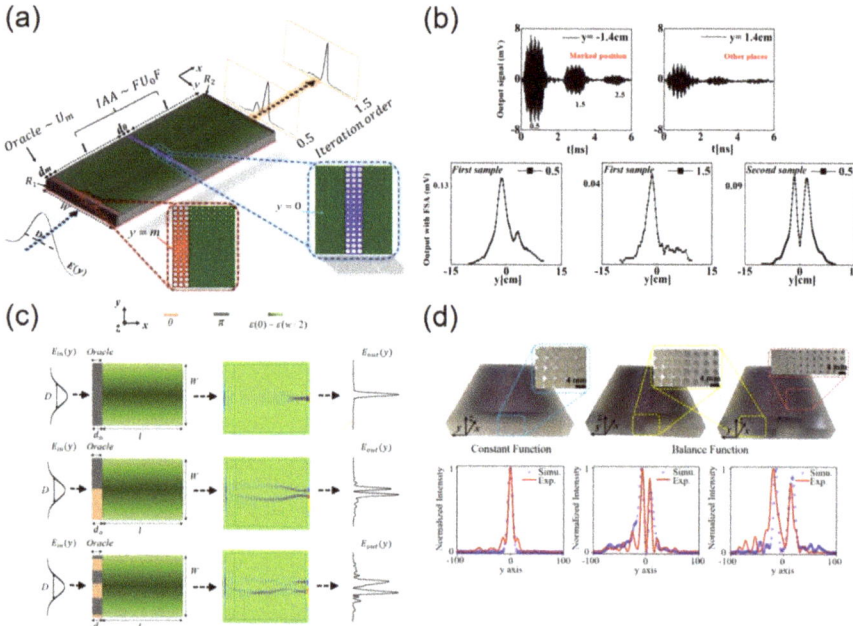

Figure 13. (a) Schematic of the general protocol of simulating quantum search algorithm with metamaterials. (b) Experimental results of the time traces (top) and the field intensity (bottom) of the output with different iterations. Adapted with permission from [154], Wiley, 2017. (c) Simulation results of metamaterial-based DJ algorithm analogy with constant, simple balance and complex balance functions, respectively. (d) The photograph (top) and experimental results of three samples for simulating DJ algorithm. Adapted with permission from [155] © The Optical Society.

Table 1. The general protocol of performing Grover's search algorithm with metamaterial in quantum and classical realm.

	Quantum	Classical
Items in the database $\|i\rangle$		"y"
Probability amplitude of the equivalent quantum state		"E(y)"
The maximum number of the database N		"D"
U_m	$I - 2\|s\rangle\langle s\|$	$\exp(i2\pi\sqrt{\varepsilon_m(y)}d_m/\lambda_0)$
	$H^{\otimes n}$	F
IAA	$I - 2\|0\rangle\langle 0\|$	$\exp(i2\pi\sqrt{\varepsilon_0(y)}d_0/\lambda_0)$
	$H^{\otimes n}$	F

In Table 1, we compare the general protocol of performing Grover's search algorithm in quantum and classical realm to clarify the differences between the two strategies. As the spatial freedom of the optical field (cbit) is adopted to simulate the qubit, the classical wave-based architecture shows a simple and enlightening way to the field of signal processing. It should be mentioned that despite that it has been proved that entanglement is not necessary for the efficiency of some quantum algorithms [156], a lack of entanglement will essentially limit the physical resources for these classical analogies, leading to an exponential growth of the width of the beam as the numbers of qubits increase. In addition, the size of the database is also limited by the spatial resolution, diffraction effect and paraxial approximation of the optical system. With the rapid development of the field of artificial electromagnetism, we can expect more novel functionalities on simulating quantum behaviors, such as three-dimensional all-optical search and data classification.

4. Conclusions and Outlook

In this review, we have presented some of the most exciting developments in the field of analogous optical computing taking advantage of metamaterials, offering an integrated and miniaturized platform for image processing and edge detection to overcome the limitations of digital electronic computers. With continuous exploitation, tremendous novel approaches have been proposed based on different physical mechanisms, including Brewster effect, surface plasmonic, photonic spin Hall effect, local/excitation mode coupling to implement mathematical operations. Despite the promising unprecedent advances of meta-structures over traditional optical system for analog computing, to date, it is still extremely challenging to further improve the performance which is limited by spatial resolution, efficiency, operating bandwidth, strong polarization/angle/symmetry dependence. In general, computational metamaterials can be divided into two major categories: plasmonics and dielectrics. Leveraging highly confined surface plasmonic modes, plasmonic spatial differentiators process advantages of simplicity and ultracompact size. However, they suffer from some fundamental limitations, for instance, the critical coupling condition limits the incident angles, the presence of higher-order Taylor coefficients lead to a narrow spatial bandwidth and further effect the resolution of the computational system. In addition, the conversion efficiency is restricted by intrinsic material loss. To improve the efficiency and achieve high-resolution edge detection, all-dielectric metasurface is emerging as an alternative platform, which is a promising candidate for further broadening of the operational spatial bandwidth and reducing the absorption losses. It should be mentioned that dielectric metasurfaces also have some drawbacks—some of them only work for one polarization while the energy of other polarized waves is completely wasted, which may lead to relatively low signal-to-noise ratio. Moreover, determining how to balance the features of numerical aperture, efficiency and spatial resolution requires an elaborate design [157].

Although some limitations exist in this emerging field, we are pleased to see that many achievements have been proposed and demonstrated to overcome these drawbacks. Wang et al. [158] proposed a photonic crystal differentiator with improved robustness with incoherent light. Kwon et al. [145] designed a nonlocal metasurface by engineering nonlocality in momentum space that can perform both even- and odd-order differentiations. This study provides a high-quality, efficient and polarization-insensitive image processing platform for 2D edge detection. We believe that as it continuously evolves, the metamaterial-based analogous optical computing will be extensively applied in signal and image processing.

Author Contributions: Y.F. conceived the review; K.C., Y.F. wrote the manuscript; W.Z., Y.G., S.F., H.L. revised the manuscript. All authors have read and agreed to the published version of the manuscript.

Funding: This research was funded by the National Natural Science Foundation of China (No. 11774057, 12074314, 11674266), This research was funded by Funds for Equipment Advance Research during the 13th Five-Year Plan under Grant 61400030305.

Data Availability Statement: Data sharing not applicable.

Conflicts of Interest: The authors declare no conflict of interest.

References

1. Clymer, A.B. The mechanical analog computers of Hannibal Ford and William Newell. *IEEE Ann. Hist. Comput.* **1993**, *15*, 19–34. [CrossRef]
2. Price, D.D.S. A History of Calculating Machines. *IEEE Micro* **1984**, *4*, 22–52. [CrossRef]
3. Hausner, A. *Analog and Analog/Hybrid Computer Programming*; Prentice-Hall: Englewood Cliffs, NJ, USA, 1971. [CrossRef]
4. Barrios, G.A.; Retamal, J.C.; Solano, E.; Sanz, M. Analog simulator of integro-differential equations with classical memristors. *Sci. Rep.* **2019**, *9*, 12928. [CrossRef] [PubMed]
5. Tsividis, Y. Not your Father's analog computer. *IEEE Spectr.* **2018**, *55*, 38–43. [CrossRef]
6. Caulfield, H.J.; Dolev, S. Why future supercomputing requires optics. *Nat. Photonics* **2010**, *4*, 261–263. [CrossRef]
7. The power of analogies. *Nat. Photonics* **2014**, *8*. [CrossRef]

8. Passian, A.; Imam, N. Nanosystems, Edge Computing, and the Next Generation Computing Systems. *Sensors* **2019**, *19*, 4048. [CrossRef]
9. Solli, D.R.; Jalali, B. Analog optical computing. *Nat. Photonics* **2015**, *9*, 704–706. [CrossRef]
10. Karim, M.A.; Awwal, A.A. *Optical Computing: An Introduction*; Wiley, Inc.: New York, NY, USA, 1992.
11. Herden, A.; Tschudi, T. *Analog Optical Computing*; Springer: Berlin/Heidelberg, Germany, 1986; pp. 369–383.
12. Creasey, D.J. Digital Signal Processing: Principles, Devices and Applications. *IEE Rev.* **1990**, *36*, 275–276. [CrossRef]
13. Park, Y.; Ahn, T.-J.; Dai, Y.; Yao, J.; Azaña, J. All-optical temporal integration of ultrafast pulse waveforms. *Opt. Express* **2008**, *16*, 17817–17825. [CrossRef]
14. Quoc Ngo, N. Design of an optical temporal integrator based on a phase-shifted fiber Bragg grating in transmission. *Opt. Lett.* **2007**, *32*, 3020–3022. [CrossRef] [PubMed]
15. Rutkowska, K.A.; Duchesne, D.; Strain, M.J.; Morandotti, R.; Sorel, M.; Azaña, J. Ultrafast all-optical temporal differentiators based on CMOS-compatible integrated-waveguide Bragg gratings. *Opt. Express* **2011**, *19*, 19514–19522. [CrossRef] [PubMed]
16. Slavík, R.; Park, Y.; Kulishov, M.; Morandotti, R.; Azaña, J. Ultrafast all-optical differentiators. *Opt. Express* **2006**, *14*, 10699–10707. [CrossRef] [PubMed]
17. Xu, J.; Zhang, X.; Dong, J.; Liu, D.; Huang, D. High-speed all-optical differentiator based on a semiconductor optical amplifier and an optical filter. *Opt. Lett.* **2007**, *32*, 1872–1874. [CrossRef] [PubMed]
18. Ferrera, M.; Park, Y.; Razzari, L.; Little, B.E.; Chu, S.T.; Morandotti, R.; Moss, D.J.; Azaña, J. On-chip CMOS-compatible all-optical integrator. *Nat. Commun.* **2010**, *1*, 29. [CrossRef] [PubMed]
19. Zangeneh-Nejad, F.; Fleury, R. Topological analog signal processing. *Nat. Commun.* **2019**, *10*, 2058. [CrossRef]
20. Chen, K.; Hou, J.; Huang, Z.; Cao, T.; Zhang, J.; Yu, Y.; Zhang, X. All-optical 1st- and 2nd-order differential equation solvers with large tuning ranges using Fabry-Pérot semiconductor optical amplifiers. *Opt. Express* **2015**, *23*, 3784–3794. [CrossRef]
21. Yang, T.; Dong, J.; Lu, L.; Zhou, L.; Zheng, A.; Zhang, X.; Chen, J. All-optical differential equation solver with constant-coefficient tunable based on a single microring resonator. *Sci. Rep.* **2014**, *4*, 5581. [CrossRef]
22. Karimi, A.; Zarifkar, A.; Miri, M. Design of a Miniaturized Broadband Silicon Hybrid Plasmonic Temporal Integrator for Ultrafast Optical Signal Processing. *J. Lightwave Technol.* **2020**, *38*, 2346–2352. [CrossRef]
23. Wu, J.; Liu, B.; Peng, J.; Mao, J.; Jiang, X.; Qiu, C.; Tremblay, C.; Su, Y. On-Chip Tunable Second-Order Differential-Equation Solver Based on a Silicon Photonic Mode-Split Microresonator. *J. Lightwave Technol.* **2015**, *33*, 3542–3549. [CrossRef]
24. Zhang, W.; Yao, J. Photonic integrated field-programmable disk array signal processor. *Nat. Commun.* **2020**, *11*, 406. [CrossRef] [PubMed]
25. Zhang, J.; Guzzon, R.S.; Coldren, L.A.; Yao, J. Optical dynamic memory based on an integrated active ring resonator. *Opt. Lett.* **2018**, *43*, 4687–4690. [CrossRef] [PubMed]
26. Brunner, D.; Soriano, M.C.; Mirasso, C.R.; Fischer, I. Parallel photonic information processing at gigabyte per second data rates using transient states. *Nat. Commun.* **2013**, *4*, 1364. [CrossRef]
27. Moughames, J.; Porte, X.; Thiel, M.; Ulliac, G.; Larger, L.; Jacquot, M.; Kadic, M.; Brunner, D. Three-dimensional waveguide interconnects for scalable integration of photonic neural networks. *Optica* **2020**, *7*, 640–646. [CrossRef]
28. Solli, D.R.; Herink, G.; Jalali, B.; Ropers, C. Fluctuations and correlations in modulation instability. *Nat. Photonics* **2012**, *6*, 463–468. [CrossRef]
29. Donati, S.; Mirasso, C.R. Introduction to the feature section on optical chaos and applications to cryptography. *IEEE J. Quantum Electron.* **2002**, *38*, 1138–1140. [CrossRef]
30. Turitsyna, E.G.; Smirnov, S.V.; Sugavanam, S.; Tarasov, N.; Shu, X.; Babin, S.A.; Podivilov, E.V.; Churkin, D.V.; Falkovich, G.; Turitsyn, S.K. The laminar–turbulent transition in a fibre laser. *Nat. Photonics* **2013**, *7*, 783–786. [CrossRef]
31. Bykov, D.A.; Doskolovich, L.L.; Bezus, E.A.; Soifer, V.A. Optical computation of the Laplace operator using phase-shifted Bragg grating. *Opt. Express* **2014**, *22*, 25084–25092. [CrossRef]
32. AbdollahRamezani, S.; Arik, K.; Khavasi, A.; Kavehvash, Z. Analog computing using graphene-based metalines. *Opt. Lett.* **2015**, *40*, 5239–5242. [CrossRef]
33. Zangeneh-Nejad, F.; Khavasi, A.; Rejaei, B. Analog optical computing by half-wavelength slabs. *Opt. Commun.* **2018**, *407*, 338–343. [CrossRef]
34. Shor, P.W. Algorithms for quantum computation: Discrete logarithms and factoring. In Proceedings of the 35th Annual Symposium on Foundations of Computer Science, Los Alamitos, CA, USA, 20–22 November 1994.
35. Grover, L.K. A fast quantum mechanical algorithm for database search. In Proceedings of the 28th Annual ACM Symposium on Theory of Computing (STOC), Philadelphia, PA, USA, 6–8 July 1996; pp. 212–219.
36. Grover, L.K. Quantum Computers Can Search Arbitrarily Large Databases by a Single Query. *Phys. Rev. Lett.* **1997**, *79*, 4709–4712. [CrossRef]
37. Knill, E.; Laflamme, R.; Milburn, G.J. A scheme for efficient quantum computation with linear optics. *Nature* **2001**, *409*, 46–52. [CrossRef] [PubMed]
38. Nielsen, M.A.; Chuang, I.L. *Quantum Computation and Quantum Information*; Cambridge University Press: New York, NY, USA, 2000.
39. Spreeuw, R.J.C. A Classical Analogy of Entanglement. *Found. Phys.* **1998**, *28*, 361–374. [CrossRef]
40. Spreeuw, R.J.C. Classical wave-optics analogy of quantum-information processing. *Phys. Rev. A* **2001**, *63*, 062302. [CrossRef]
41. Collins, D.; Kim, K.W.; Holton, W.C. Deutsch-Jozsa algorithm as a test of quantum computation. *Phys. Rev. A* **1998**, *58*, R1633–R1636. [CrossRef]

42. Kwiat, P.G.; Mitchell, J.R.; Schwindt, P.D.D.; White, A.G. Grover's search algorithm: An optical approach. *J. Mod. Opt.* **2000**, *47*, 257–266. [CrossRef]
43. Bhattacharya, N.; van Linden van den Heuvell, H.B.; Spreeuw, R.J.C. Implementation of Quantum Search Algorithm using Classical Fourier Optics. *Phys. Rev. Lett.* **2002**, *88*, 137901. [CrossRef]
44. Dragoman, D. n-step optical simulation of the n-qubit state: Applications in optical computing. *Optik* **2002**, *113*, 425–428. [CrossRef]
45. Puentes, G.; Mela, C.L.; Ledesma, S.; Iemmi, C.; Paz, J.P.; Saraceno, M. Optical simulation of quantum algorithms using programmable liquid-crystal displays. *Phys. Rev. A* **2004**, *69*, 042319. [CrossRef]
46. Hijmans, T.W.; Huussen, T.N.; Spreeuw, R.J.C. Time and frequency domain solutions in an optical analogue of Grover's search algorithm. *J. Opt. Soc. Am. B* **2006**, *24*, 214–220. [CrossRef]
47. Luis, A. Coherence, polarization, and entanglement for classical light fields. *Opt. Commun.* **2009**, *282*, 3665–3670. [CrossRef]
48. Zhang, P.; Liu, R.-F.; Huang, Y.-F.; Gao, H.; Li, F.-L. Demonstration of Deutsch's algorithm on a stable linear optical quantum computer. *Phys. Rev. A* **2010**, *82*, 064302. [CrossRef]
49. Lee, S.M.; Park, H.S.; Cho, J.; Kang, Y.; Lee, J.Y.; Kim, H.; Lee, D.-H.; Choi, S.-K. Experimental realization of a four-photon seven-qubit graph state for one-way quantum computation. *Opt. Express* **2012**, *20*, 6915–6926. [CrossRef] [PubMed]
50. Marques, B.; Barros, M.R.; Pimenta, W.M.; Carvalho, M.A.D.; Ferraz, J.; Drumond, R.C.; Terra Cunha, M.; Pádua, S. Double-slit implementation of the minimal Deutsch algorithm. *Phys. Rev. A* **2012**, *86*, 032306. [CrossRef]
51. Rohde, P.P. Optical quantum computing with photons of arbitrarily low fidelity and purity. *Phys. Rev. A* **2012**, *86*, 052321. [CrossRef]
52. Goyal, S.K.; Roux, F.S.; Forbes, A.; Konrad, T. Implementing Quantum Walks Using Orbital Angular Momentum of Classical Light. *Phys. Rev. Lett.* **2013**, *110*, 263602. [CrossRef]
53. Georgescu, I.M.; Ashhab, S.; Nori, F. Quantum simulation. *Rev. Mod. Phys.* **2014**, *86*, 153–185. [CrossRef]
54. Goyal, S.K.; Roux, F.S.; Forbes, A.; Konrad, T. The Scalable Implementation of Quantum Walks using Classical Light. In Proceedings of the SPIE OPTO, San Francisco, CA, USA, 1–6 February 2014; Volume 8999, p. 8.
55. Böhm, J.; Bellec, M.; Mortessagne, F.; Kuhl, U.; Barkhofen, S.; Gehler, S.; Stöckmann, H.-J.; Foulger, I.; Gnutzmann, S.; Tanner, G. Microwave Experiments Simulating Quantum Search and Directed Transport in Artificial Graphene. *Phys. Rev. Lett.* **2015**, *114*, 110501. [CrossRef]
56. Goyal, S.K.; Roux, F.S.; Forbes, A.; Konrad, T. Implementation of multidimensional quantum walks using linear optics and classical light. *Phys. Rev. A* **2015**, *92*, 040302. [CrossRef]
57. Hor-Meyll, M.; Tasca, D.S.; Walborn, S.P.; Ribeiro, P.H.S.; Santos, M.M.; Duzzioni, E.I. Deterministic quantum computation with one photonic qubit. *Phys. Rev. A* **2015**, *92*, 012337. [CrossRef]
58. Perez-Garcia, B.; Francis, J.; McLaren, M.; Hernandez-Aranda, R.I.; Forbes, A.; Konrad, T. Quantum computation with classical light: The Deutsch Algorithm. *Phys. Lett. A* **2015**, *379*, 1675–1680. [CrossRef]
59. Perez-Garcia, B.; Mclaren, M.; Goyal, S.K.; Hernandez-Aranda, R.I.; Forbes, A.; Konrad, T. Quantum computation with classical light: Implementation of the Deutsch–Jozsa algorithm. *Phys. Lett. A* **2015**, *380*, 1925–1931. [CrossRef]
60. Zagoskin, A.M.; Felbacq, D.; Rousseau, E. Quantum metamaterials in the microwave and optical ranges. *EPJ Quantum Technol.* **2016**, *3*, 2. [CrossRef]
61. Giri, P.R.; Korepin, V.E. A review on quantum search algorithms. *Quantum Inf. Process.* **2017**, *16*, 315. [CrossRef]
62. Johansson, N.; Larsson, J.-Å. Efficient classical simulation of the Deutsch–Jozsa and Simon's algorithms. *Quantum Inf. Process.* **2017**, *16*, 233. [CrossRef]
63. Wei, H.-R.; Liu, J.-Z. Deterministic implementations of single-photon multi-qubit Deutsch–Jozsa algorithms with linear optics. *Annals Phys.* **2017**, *377*, 38–47. [CrossRef]
64. Flamini, F.; Spagnolo, N.; Sciarrino, F. Photonic quantum information processing: A review. *Rep. Prog. Phys.* **2018**, *82*, 016001. [CrossRef] [PubMed]
65. Perez-Garcia, B.; Hernandez-Aranda, R.I.; Forbes, A.; Konrad, T. The first iteration of Grover's algorithm using classical light with orbital angular momentum. *J. Mod. Opt.* **2018**, *65*, 1942–1948. [CrossRef]
66. Vianna, Y.; Barros, M.R.; Hor-Meyll, M. Classical realization of the quantum Deutsch algorithm. *Am. J. Phys.* **2018**, *86*, 914–923. [CrossRef]
67. Zhang, S.; Li, P.; Wang, B.; Zeng, Q.; Zhang, X. Implementation of quantum permutation algorithm with classical light. *J. Phys. Commun.* **2019**, *3*, 015008. [CrossRef]
68. Zhang, S.; Zhang, Y.; Sun, Y.; Sun, H.; Zhang, X. Quantum-inspired microwave signal processing for implementing unitary transforms. *Opt. Express* **2019**, *27*, 436–460. [CrossRef] [PubMed]
69. Smith, D.R.; Pendry, J.B.; Wiltshire, M.C.K. Metamaterials and Negative Refractive Index. *Science* **2004**, *305*, 788–792. [CrossRef] [PubMed]
70. Chen, H.; Chan, C.T.; Sheng, P. Transformation optics and metamaterials. *Nat. Mater.* **2010**, *9*, 387. [CrossRef] [PubMed]
71. Liu, Y.; Zhang, X. Metamaterials: A new frontier of science and technology. *Chem. Soc. Rev.* **2011**, *40*, 2494–2507. [CrossRef]
72. Soukoulis, C.M.; Wegener, M. Past achievements and future challenges in the development of three-dimensional photonic metamaterials. *Nat. Photonics* **2011**, *5*, 523. [CrossRef]

73. Cui, T.J.; Qi, M.Q.; Wan, X.; Zhao, J.; Cheng, Q. Coding metamaterials, digital metamaterials and programmable metamaterials. *Light Sci. Appl.* **2014**, *3*, e218. [CrossRef]
74. Della Giovampaola, C.; Engheta, N. Digital metamaterials. *Nat. Mater.* **2014**, *13*, 1115. [CrossRef]
75. Huo, P.; Zhang, S.; Liang, Y.; Lu, Y.; Xu, T. Hyperbolic Metamaterials and Metasurfaces: Fundamentals and Applications. *Adv. Opt. Mater.* **2019**, *7*, 1801616. [CrossRef]
76. Fan, Y.; Wei, Z.; Li, H.; Chen, H.; Soukoulis, C.M. Low-loss and high-Q planar metamaterial with toroidal moment. *Phys. Rev. B* **2013**, *87*, 115417. [CrossRef]
77. Chen, S.; Fan, Y.; Fu, Q.; Wu, H.; Jin, Y.; Zheng, J.; Zhang, F. A Review of Tunable Acoustic Metamaterials. *Appl. Sci.* **2018**, *8*, 1480. [CrossRef]
78. Fan, Y.; Zhang, F.; Shen, N.-H.; Fu, Q.; Wei, Z.; Li, H.; Soukoulis, C.M. Achieving a high-Q response in metamaterials by manipulating the toroidal excitations. *Phys. Rev. A* **2018**, *97*, 033816. [CrossRef]
79. Ziolkowski, R.W.; Engheta, N. Metamaterials: Two Decades Past and Into Their Electromagnetics Future and Beyond. *IEEE Trans. Antennas Propag.* **2020**, *68*, 1232–1237. [CrossRef]
80. Yu, N.; Genevet, P.; Kats, M.A.; Aieta, F.; Tetienne, J.-P.; Capasso, F.; Gaburro, Z. Light Propagation with Phase Discontinuities: Generalized Laws of Reflection and Refraction. *Science* **2011**, *334*, 333–337. [CrossRef] [PubMed]
81. Holloway, C.L.; Kuester, E.F.; Gordon, J.A.; Hara, J.O.; Booth, J.; Smith, D.R. An Overview of the Theory and Applications of Metasurfaces: The Two-Dimensional Equivalents of Metamaterials. *IEEE Antennas Propag. Mag.* **2012**, *54*, 10–35. [CrossRef]
82. Kildishev, A.V.; Boltasseva, A.; Shalaev, V.M. Planar Photonics with Metasurfaces. *Science* **2013**, *339*, 1232009. [CrossRef]
83. Meinzer, N.; Barnes, W.L.; Hooper, I.R. Plasmonic meta-atoms and metasurfaces. *Nat. Photonics* **2014**, *8*, 889–898. [CrossRef]
84. Yu, N.; Capasso, F. Flat optics with designer metasurfaces. *Nat. Mater.* **2014**, *13*, 139. [CrossRef]
85. Glybovski, S.B.; Tretyakov, S.A.; Belov, P.A.; Kivshar, Y.S.; Simovski, C.R. Metasurfaces: From microwaves to visible. *Phys. Rep.* **2016**, *634*, 1–72. [CrossRef]
86. Hou-Tong, C.; Antoinette, J.T.; Nanfang, Y. A review of metasurfaces: Physics and applications. *Rep. Prog. Phys.* **2016**, *79*, 076401. [CrossRef]
87. Zhang, L.; Mei, S.; Huang, K.; Qiu, C.-W. Advances in Full Control of Electromagnetic Waves with Metasurfaces. *Adv. Opt. Mater.* **2016**, *4*, 818–833. [CrossRef]
88. Genevet, P.; Capasso, F.; Aieta, F.; Khorasaninejad, M.; Devlin, R. Recent advances in planar optics: From plasmonic to dielectric metasurfaces. *Optica* **2017**, *4*, 139–152. [CrossRef]
89. Li, G.; Zhang, S.; Zentgraf, T. Nonlinear photonic metasurfaces. *Nat. Rev. Mater.* **2017**, *2*, 17010. [CrossRef]
90. Chen, M.; Kim, M.; Wong Alex, M.H.; Eleftheriades George, V. Huygens' metasurfaces from microwaves to optics: A review. *Nanophotonics* **2018**, *7*, 1207. [CrossRef]
91. He, Q.; Sun, S.; Xiao, S.; Zhou, L. High-Efficiency Metasurfaces: Principles, Realizations, and Applications. *Adv. Opt. Mater.* **2018**, *6*, 1800415. [CrossRef]
92. Hail, C.U.; Michel, A.-K.U.; Poulikakos, D.; Eghlidi, H. Optical Metasurfaces: Evolving from Passive to Adaptive. *Adv. Opt. Mater.* **2019**, *7*, 1801786. [CrossRef]
93. Luo, X. Metamaterials and Metasurfaces. *Adv. Opt. Mater.* **2019**, *7*, 1900885. [CrossRef]
94. Sun, S.; He, Q.; Hao, J.; Xiao, S.; Zhou, L. Electromagnetic metasurfaces: Physics and applications. *Adv. Opt. Photonics* **2019**, *11*, 380–479. [CrossRef]
95. Fan, Y.; Shen, N.-H.; Zhang, F.; Zhao, Q.; Wei, Z.; Zhang, P.; Dong, J.; Fu, Q.; Li, H.; Soukoulis, C.M. Photoexcited Graphene Metasurfaces: Significantly Enhanced and Tunable Magnetic Resonances. *ACS Photonics* **2018**, *5*, 1612–1618. [CrossRef]
96. Fan, Y.; Shen, N.-H.; Zhang, F.; Zhao, Q.; Wu, H.; Fu, Q.; Wei, Z.; Li, H.; Soukoulis, C.M. Graphene Plasmonics: A Platform for 2D Optics. *Adv. Opt. Mater.* **2018**, *7*, 1800537. [CrossRef]
97. Cheng, K.; Wei, Z.; Fan, Y.; Zhang, X.; Wu, C.; Li, H. Realizing Broadband Transparency via Manipulating the Hybrid Coupling Modes in Metasurfaces for High-Efficiency Metalens. *Adv. Opt. Mater.* **2019**, *7*, 1900016. [CrossRef]
98. Lou, J.; Liang, J.; Yu, Y.; Ma, H.; Yang, R.; Fan, Y.; Wang, G.; Cai, T. Silicon-Based Terahertz Meta-Devices for Electrical Modulation of Fano Resonance and Transmission Amplitude. *Adv. Opt. Mater.* **2020**, *8*, 2000449. [CrossRef]
99. Zhu, W.; Fan, Y.; Li, C.; Yang, R.; Yan, S.; Fu, Q.; Zhang, F.; Gu, C.; Li, J. Realization of a near-infrared active Fano-resonant asymmetric metasurface by precisely controlling the phase transition of Ge2Sb2Te5. *Nanoscale* **2020**, *12*, 8758–8767. [CrossRef] [PubMed]
100. Wang, Z.; Li, T.; Soman, A.; Mao, D.; Kananen, T.; Gu, T. On-chip wavefront shaping with dielectric metasurface. *Nat. Commun.* **2019**, *10*, 3547. [CrossRef] [PubMed]
101. Abdollahramezani, S.; Hemmatyar, O.; Adibi, A. Meta-optics for spatial optical analog computing. *Nanophotonics* **2020**, *9*, 4075–4095. [CrossRef]
102. Silva, A.; Monticone, F.; Castaldi, G.; Galdi, V.; Alù, A.; Engheta, N. Performing Mathematical Operations with Metamaterials. *Science* **2014**, *343*, 160. [CrossRef]
103. Mendlovic, D.; Ozaktas, H.M. Fractional Fourier transforms and their optical implementation: I. *J. Opt. Soc. Am. A* **1993**, *10*, 1875–1881. [CrossRef]
104. Ozaktas, H.M.; Mendlovic, D. Fractional Fourier transforms and their optical implementation. II. *J. Opt. Soc. Am. A* **1993**, *10*, 2522–2531. [CrossRef]

105. Khorasaninejad, M.; Capasso, F. Metalenses: Versatile multifunctional photonic components. *Science* **2017**, *358*, eaam8100. [CrossRef]
106. Lalanne, P.; Chavel, P. Metalenses at visible wavelengths: Past, present, perspectives. *Laser Photonics Rev.* **2017**, *11*, 1600295. [CrossRef]
107. Wang, S.; Wu, P.C.; Su, V.-C.; Lai, Y.-C.; Hung Chu, C.; Chen, J.-W.; Lu, S.-H.; Chen, J.; Xu, B.; Kuan, C.-H.; et al. Broadband achromatic optical metasurface devices. *Nat. Commun.* **2017**, *8*, 187. [CrossRef]
108. Yang, Q.; Gu, J.; Xu, Y.; Zhang, X.; Li, Y.; Ouyang, C.; Tian, Z.; Han, J.; Zhang, W. Broadband and Robust Metalens with Nonlinear Phase Profiles for Efficient Terahertz Wave Control. *Adv. Opt. Mater.* **2017**, *5*, 1601084. [CrossRef]
109. Chen, W.T.; Zhu, A.Y.; Sanjeev, V.; Khorasaninejad, M.; Shi, Z.; Lee, E.; Capasso, F. A broadband achromatic metalens for focusing and imaging in the visible. *Nat. Nanotechnol.* **2018**, *13*, 220–226. [CrossRef] [PubMed]
110. Shrestha, S.; Overvig, A.C.; Lu, M.; Stein, A.; Yu, N. Broadband achromatic dielectric metalenses. *Light Sci. Appl.* **2018**, *7*, 85. [CrossRef] [PubMed]
111. Huang, L.; Chen, X.; Mühlenbernd, H.; Zhang, H.; Chen, S.; Bai, B.; Tan, Q.; Jin, G.; Cheah, K.-W.; Qiu, C.-W.; et al. Three-dimensional optical holography using a plasmonic metasurface. *Nat. Commun.* **2013**, *4*. [CrossRef]
112. Li, G.; Kang, M.; Chen, S.; Zhang, S.; Pun, E.Y.-B.; Cheah, K.W.; Li, J. Spin-Enabled Plasmonic Metasurfaces for Manipulating Orbital Angular Momentum of Light. *Nano Lett.* **2013**, *13*, 4148–4151. [CrossRef]
113. Lovera, A.; Gallinet, B.; Nordlander, P.; Martin, O.J.F. Mechanisms of Fano Resonances in Coupled Plasmonic Systems. *ACS Nano* **2013**, *7*, 4527–4536. [CrossRef]
114. Ni, X.; Ishii, S.; Kildishev, A.V.; Shalaev, V.M. Ultra-thin, planar, Babinet-inverted plasmonic metalenses. *Light Sci. Appl.* **2013**, *2*, e72. [CrossRef]
115. Pors, A.; Nielsen, M.G.; Eriksen, R.L.; Bozhevolnyi, S.I. Broadband Focusing Flat Mirrors Based on Plasmonic Gradient Metasurfaces. *Nano Lett.* **2013**, *13*, 829–834. [CrossRef]
116. Chalabi, H.; Schoen, D.; Brongersma, M.L. Hot-Electron Photodetection with a Plasmonic Nanostripe Antenna. *Nano Lett.* **2014**, *14*, 1374–1380. [CrossRef]
117. Saeidi, C.; van der Weide, D. Wideband plasmonic focusing metasurfaces. *Appl. Phys. Lett.* **2014**, *105*, 053107. [CrossRef]
118. Qin, F.; Ding, L.; Zhang, L.; Monticone, F.; Chum, C.C.; Deng, J.; Mei, S.; Li, Y.; Teng, J.; Hong, M.; et al. Hybrid bilayer plasmonic metasurface efficiently manipulates visible light. *Sci. Adv.* **2016**, *2*. [CrossRef] [PubMed]
119. Yue, F.; Wen, D.; Xin, J.; Gerardot, B.D.; Li, J.; Chen, X. Vector Vortex Beam Generation with a Single Plasmonic Metasurface. *ACS Photonics* **2016**, *3*, 1558–1563. [CrossRef]
120. Pors, A.; Nielsen, M.G.; Bozhevolnyi, S.I. Analog Computing Using Reflective Plasmonic Metasurfaces. *Nano Lett.* **2015**, *15*, 791–797. [CrossRef] [PubMed]
121. Chen, H.; An, D.; Li, Z.; Zhao, X. Performing differential operation with a silver dendritic metasurface at visible wavelengths. *Opt. Express* **2017**, *25*, 26417–26426. [CrossRef]
122. Chizari, A.; Abdollahramezani, S.; Jamali, M.V.; Salehi, J.A. Analog optical computing based on a dielectric meta-reflect array. *Opt. Lett.* **2016**, *41*, 3451–3454. [CrossRef]
123. Zhang, W.; Qu, C.; Zhang, X. Solving constant-coefficient differential equations with dielectric metamaterials. *J. Opt.* **2016**, *18*, 075102. [CrossRef]
124. Wu, Y.; Zhuang, Z.; Deng, L.; Liu, Y.; Xue, Q.; Ghassemlooy, Z. Arbitrary Multi-way Parallel Mathematical Operations Based on Planar Discrete Metamaterials. *Plasmonics* **2018**, *13*, 599–607. [CrossRef]
125. Huo, P.; Zhang, C.; Zhu, W.; Liu, M.; Zhang, S.; Zhang, S.; Chen, L.; Lezec, H.J.; Agrawal, A.; Lu, Y.; et al. Photonic Spin-Multiplexing Metasurface for Switchable Spiral Phase Contrast Imaging. *Nano Lett.* **2020**, *20*, 2791–2798. [CrossRef]
126. Wu, W.; Jiang, W.; Yang, J.; Gong, S.; Ma, Y. Multilayered analog optical differentiating device: Performance analysis on structural parameters. *Opt. Lett.* **2017**, *42*, 5270–5273. [CrossRef]
127. Zangeneh-Nejad, F.; Khavasi, A. Spatial integration by a dielectric slab and its planar graphene-based counterpart. *Opt. Lett.* **2017**, *42*, 1954–1957. [CrossRef]
128. Guo, C.; Xiao, M.; Minkov, M.; Shi, Y.; Fan, S. Photonic crystal slab Laplace operator for image differentiation. *Optica* **2018**, *5*, 251–256. [CrossRef]
129. Doskolovich, L.L.; Bezus, E.A.; Bykov, D.A.; Soifer, V.A. Spatial differentiation of Bloch surface wave beams using an on-chip phase-shifted Bragg grating. *J. Opt.* **2016**, *18*, 115006. [CrossRef]
130. Fang, Y.; Lou, Y.; Ruan, Z. On-grating graphene surface plasmons enabling spatial differentiation in the terahertz region. *Opt. Lett.* **2017**, *42*, 3840–3843. [CrossRef]
131. Bykov, D.A.; Doskolovich, L.L.; Morozov, A.A.; Podlipnov, V.V.; Bezus, E.A.; Verma, P.; Soifer, V.A. First-order optical spatial differentiator based on a guided-mode resonant grating. *Opt. Express* **2018**, *26*, 10997–11006. [CrossRef] [PubMed]
132. Dong, Z.; Si, J.; Yu, X.; Deng, X. Optical spatial differentiator based on subwavelength high-contrast gratings. *Appl. Phys. Lett.* **2018**, *112*, 181102. [CrossRef]
133. Yang, W.; Yu, X.; Zhang, J.; Deng, X. Plasmonic transmitted optical differentiator based on the subwavelength gold gratings. *Opt. Lett.* **2020**, *45*, 2295–2298. [CrossRef]
134. Parthenopoulos, A.; Darki, A.A.; Jeppesen, B.R.; Dantan, A. Optical spatial differentiation with suspended subwavelength gratings. *arXiv* **2020**, arXiv:2008.10945.
135. Zhu, T.; Zhou, Y.; Lou, Y.; Ye, H.; Qiu, M.; Ruan, Z.; Fan, S. Plasmonic computing of spatial differentiation. *Nat. Commun.* **2017**, *8*, 15391. [CrossRef]

136. Zhang, J.; Ying, Q.; Ruan, Z. Time response of plasmonic spatial differentiators. *Opt. Lett.* **2019**, *44*, 4511–4514. [CrossRef]
137. Zhang, W.; Zhang, X. Backscattering-Immune Computing of Spatial Differentiation by Nonreciprocal Plasmonics. *Phys. Rev. Appl.* **2019**, *11*, 054033. [CrossRef]
138. Cordaro, A.; Kwon, H.; Sounas, D.; Koenderink, A.F.; Alù, A.; Polman, A. High-Index Dielectric Metasurfaces Performing Mathematical Operations. *Nano Lett.* **2019**, *19*, 8418–8423. [CrossRef] [PubMed]
139. Zhou, Y.; Wu, W.; Chen, R.; Chen, W.; Chen, R.; Ma, Y. Analog Optical Spatial Differentiators Based on Dielectric Metasurfaces. *Adv. Opt. Mater.* **2020**, *8*, 1901523. [CrossRef]
140. Zhou, J.; Qian, H.; Chen, C.-F.; Zhao, J.; Li, G.; Wu, Q.; Luo, H.; Wen, S.; Liu, Z. Optical edge detection based on high-efficiency dielectric metasurface. *Proc. Natl. Acad. Sci. USA* **2019**, *116*, 11137. [CrossRef] [PubMed]
141. Wan, L.; Pan, D.; Yang, S.; Zhang, W.; Potapov, A.A.; Wu, X.; Liu, W.; Feng, T.; Li, Z. Optical analog computing of spatial differentiation and edge detection with dielectric metasurfaces. *Opt. Lett.* **2020**, *45*, 2070–2073. [CrossRef] [PubMed]
142. Zhou, Y.; Zheng, H.; Kravchenko, I.I.; Valentine, J. Flat optics for image differentiation. *Nat. Photonics* **2020**, *14*, 316–323. [CrossRef]
143. Saba, A.; Tavakol, M.R.; Karimi-Khoozani, P.; Khavasi, A. Two-Dimensional Edge Detection by Guided Mode Resonant Metasurface. *IEEE Photonics Technol. Lett.* **2018**, *30*, 853–856. [CrossRef]
144. Kwon, H.; Sounas, D.; Cordaro, A.; Polman, A.; Alù, A. Nonlocal Metasurfaces for Optical Signal Processing. *Phys. Rev. Lett.* **2018**, *121*, 173004. [CrossRef]
145. Kwon, H.; Cordaro, A.; Sounas, D.; Polman, A.; Alù, A. Dual-Polarization Analog 2D Image Processing with Nonlocal Metasurfaces. *ACS Photonics* **2020**, *7*, 1799–1805. [CrossRef]
146. Del Hougne, P.; Lerosey, G. Leveraging Chaos for Wave-Based Analog Computation: Demonstration with Indoor Wireless Communication Signals. *Phys. Rev. X* **2018**, *8*, 041037. [CrossRef]
147. Mohammadi Estakhri, N.; Edwards, B.; Engheta, N. Inverse-designed metastructures that solve equations. *Science* **2019**, *363*, 1333. [CrossRef]
148. Youssefi, A.; Zangeneh-Nejad, F.; Abdollahramezani, S.; Khavasi, A. Analog computing by Brewster effect. *Opt. Lett.* **2016**, *41*, 3467–3470. [CrossRef] [PubMed]
149. Xu, D.; He, S.; Zhou, J.; Chen, S.; Wen, S.; Luo, H. Goos–Hänchen effect enabled optical differential operation and image edge detection. *Appl. Phys. Lett.* **2020**, *116*, 211103. [CrossRef]
150. Zhu, T.; Lou, Y.; Zhou, Y.; Zhang, J.; Huang, J.; Li, Y.; Luo, H.; Wen, S.; Zhu, S.; Gong, Q.; et al. Generalized Spatial Differentiation from the Spin Hall Effect of Light and Its Application in Image Processing of Edge Detection. *Phys. Rev. Appl.* **2019**, *11*, 034043. [CrossRef]
151. Yan, W.; Qiu, M. Two-dimensional optical edge detection. *Nat. Photonics* **2020**, *14*, 268–269. [CrossRef]
152. Monz, T.; Nigg, D.; Martinez, E.A.; Brandl, M.F.; Schindler, P.; Rines, R.; Wang, S.X.; Chuang, I.L.; Blatt, R. Realization of a scalable Shor algorithm. *Science* **2016**, *351*, 1068. [CrossRef]
153. Rakhmanov, A.L.; Zagoskin, A.M.; Savel'ev, S.; Nori, F. Quantum metamaterials: Electromagnetic waves in a Josephson qubit line. *Phys. Rev. B* **2008**, *77*, 144507. [CrossRef]
154. Zhang, W.; Cheng, K.; Wu, C.; Wang, Y.; Li, H.; Zhang, X. Implementing Quantum Search Algorithm with Metamaterials. *Adv. Mater.* **2017**, *3*, 1703986. [CrossRef]
155. Cheng, K.; Zhang, W.; Wei, Z.; Fan, Y.; Xu, C.; Wu, C.; Zhang, X.; Li, H. Simulate Deutsch-Jozsa algorithm with metamaterials. *Opt. Express* **2020**, *28*, 16230–16243. [CrossRef]
156. Lloyd, S. Quantum search without entanglement. *Phys. Rev. A* **1999**, *61*, 010301. [CrossRef]
157. Karimi, P.; Khavasi, A.; Mousavi Khaleghi, S.S. Fundamental limit for gain and resolution in analog optical edge detection. *Opt. Express* **2020**, *28*, 898–911. [CrossRef]
158. Wang, H.; Guo, C.; Zhao, Z.; Fan, S. Compact Incoherent Image Differentiation with Nanophotonic Structures. *ACS Photonics* **2020**, *7*, 338–343. [CrossRef]

MDPI
St. Alban-Anlage 66
4052 Basel
Switzerland
Tel. +41 61 683 77 34
Fax +41 61 302 89 18
www.mdpi.com

Applied Sciences Editorial Office
E-mail: applsci@mdpi.com
www.mdpi.com/journal/applsci

www.ingramcontent.com/pod-product-compliance
Lightning Source LLC
LaVergne TN
LVHW070617100526
838202LV00012B/672